COMBAT RECON

In a few minutes Major Man left his radio and called for us to join him. Using a map, he explained that Colonel Quan wanted him to have choppers land a couple of platoons beyond the Viet Cong's positions and attack them from the rear.

It was a plan for disaster. Our two platoons would be no match for a reinforced Viet Cong company. I pointed out the obvious and insisted that the only reasonable course would be to allow the platoons to set up blocking positions in the tree line on the edge of the proposed landing zone. We could then use artillery and air to force the Viet Cong to withdraw—hopefully into our block.

Major Man agreed, but said that he had already argued with Colonel Quan and had been told to "execute the plan, immediately and without change." Then, looking at me, he said, "Colonel Quan insists that an advisor accompany the platoons."

"An advisor" meant me. The hair stood up on my neck, not only because it was probably a suicide mission, but also because I sensed that Colonel Quan was trying to get me killed....

COMBAT RECON

MY YEAR WITH THE ARVN

ROBERT D. PARRISH

ST. MARTIN'S PAPERBACKS

COMBAT RECON

Copyright © 1991 by Robert D. Parrish.

Panel photograph on cover by Globe Photos; inset photograph on cover by Robert Ellison for Black Star.

All rights reserved. No part of this book may be used or reproduced in any manner whatsoever without written permission except in the case of brief quotations embodied in critical articles or reviews. For information address St. Martin's Press, 175 Fifth Avenue, New York, N.Y. 10010.

Library of Congress Catalog Card Number: 90-49212

ISBN: 0-312-92713-4

Printed in the United States of America

St. Martin's Press hardcover edition/February 1991
St. Martin's Paperbacks edition/February 1992

10 9 8 7 6 5 4 3 2 1

This book is dedicated to the 3rd Battalion, 7th Regiment; the 5th Reconnaissance Company; and the G-2 Recondo Company of the 5th ARVN Infantry Division—and *most* of the men, both Vietnamese and Americans, who served in them. *They deserved much more than we ultimately gave them.*

CONTENTS

MAPS AND FIGURES

PREFACE

I went to Vietnam in May 1967 as a twenty-six-year-old Regular Army infantry first lieutenant, as someone eager to kill Viet Cong and participate in the great adventure. I was a well-trained professional soldier, having completed the Combat Platoon Leader course (for newly commissioned infantry lieutenants), the tough Ranger course (jungle and commando training), and Airborne school (military parachuting). I had spent three years as an enlisted man in the XVIII Airborne Corps at Fort Bragg, North Carolina, and had been promoted to sergeant before returning to college and taking ROTC. Although I had only one assignment as an officer before going to Vietnam, it was as the commander of a rifle company in the 5th Mechanized Infantry Division. I also knew something about Vietnam and the Vietnamese, because the army had sent me to the Special Warfare School to learn their language and customs.

Unlike some others, I willingly went to Vietnam. In fact, I had repeatedly volunteered to go starting in 1961, when I was just a Specialist Four (corporal); this was a time when most people's understanding of the country was limited to the French involvement and Dien Bien Phu. (The army turned me down because it "had no requirement for individuals with my rank and specialty.") I had not wanted to go as an advisor in 1967, but I was unsuccessful in getting the army to change my orders to a U.S. infantry division.

I spent the first half of the year with a Vietnamese infantry battalion and the second half with a reinforced Vietnamese reconnaissance company. The entire year was filled with combat, including the 1968 Tet Offensive. Although I was wounded twice, I survived my year and returned to the U.S. to rest up for a second tour, one in which I would be commanding a rifle company in the 1st Cavalry Division—which is another story.

I remained in the army after Vietnam and had an interesting, but compared to Vietnam, unexciting career. Although I had always intended to write of my experiences in 'Nam, it wasn't until I retired (as a lieutenant colonel) that I had the time. Like most Vietnam veterans, I felt that what I had done in the two years that I'd spent "humping the bush" was important. After I was gone, I didn't want my children to look at the decorations the army gave me and say, "He got a bunch of medals in Vietnam, but we don't know what he did."

This book is not intended to be some sort of catharsis. I was more fortunate than most veterans because I stayed in the army, where I was surrounded by others who had also experienced Vietnam. We all knew what it had been like, and we spent countless hours swapping war stories. I can think of nothing worse than constantly being with people who couldn't understand what we had gone through—and really didn't care.

While the memories of Vietnam are vivid, my recollection of names is less than outstanding. I have a footlocker filled with official orders, reports, maps, and miscellaneous papers that I brought home. I used these and some photographs to put names to some of the faces I remember so well. Unfortunately, I have completely forgotten other names and have had to make up some to give the reader a better identification than "the captain" or "the colonel." In some cases I knew the designation of a battalion that we worked with, but not the company. For example, I spent a month operating with the 25th Infantry Division's 2d Battalion, 14th Infantry. It rotated the rifle companies we worked with so often that I had a hard time remembering which one was on our flank at any given moment. There is no way I could remember so many years later. I apologize to those who may read this and say, "I was with Bravo Company then, and I don't remember any of this." Well, it happened to *one* of your battalion's companies. In other cases I have changed names intentionally. (Colonel Quan, Major Humphry, Captain Jefferies, and Captain Robinson are not the real names of the individuals.) Unfortunately, not every officer, NCO, and soldier was brave, honorable, and competent—although the vast majority were all those things, or at least tried to be.

I sometimes wonder about some of the things that I did—and got away with. Then I remember that we combat unit advisors were in a unique situation. We were members of a four-man team, usually far removed from our superiors. Our lives weren't as structured as they would have been in a U.S. company or battalion. Apart from the Vietnamese with whom we fought, only our few buddies knew what we were doing. Normally the team members were officers and NCOs, and we were given little guidance on how we were to do our jobs. "Do whatever is reasonable to help your unit kill Viet Cong" was our mission, and we were left to determine how best to accomplish it. That was a heady situation for a young man, even if he was a lieutenant, captain, or midlevel noncommissioned officer.

Because of the freedom, the excitement, and our youth, and because we had grown up watching war movies, we did some incredibly wild and stupid things—and usually got away with them. When I went back to Vietnam as an infantry rifle company commander, I often longed for the "good old days." If I had tried to do what I had done as an advisor, I would have been relieved of command and probably thrown out of the army. I also might have been "fragged," because my GIs wanted a commander who would help them stay alive—not a cowboy.

This book is *not* a chronology of all the battles and firefights I was involved in during the twelve months. We averaged some sort of contact with the VC about two or three times a week. It would be impossible to describe each of them. It also would make very tedious reading, because most of the firefights followed the same pattern.

They shot at us and we shot at them. We called in artillery, helicopter gunships, airstrikes, and anything else we could get. Then they withdrew, and after evacuating our dead and wounded, we went looking for them again. Most grunts will agree that after a while it all becomes something of a blur. It was scary, miserable, and exhausting, but you become numb, and details fade from the memory. You remember the *very bad* things and the *very good* things, but much of the rest becomes just a collection of random images and feelings—definitely feelings. Combat leaves an intense impression on the mind and soul.

A number of people encouraged me to write of my experiences, especially my wife Muriel. She had heard bits and pieces of what I'd been involved in, but never the whole story. In my letters home, I seldom mentioned the fighting and focused instead on the cultural and exotic. I hadn't wanted her to worry about a green army sedan pulling up to the house, carrying an officer and a chaplain who had come to tell her that I'd been killed. It wasn't until I was wounded that she discovered that my job wasn't building schools and digging wells. When she edited the manuscript, she chewed me out whenever she read about something stupid or dangerous (mostly stupid) that I had done.

Sergeant Major Reed Jarvis and Master Sergeant (Retired) Bob Gach are two others who supported me. Reed is an avid military historian, someone with whom I tramped many battlefields, and argued how the battles should have been fought. Bob would sit for hours while I rambled on about Vietnam or anything else that happened to strike my fancy. Since he had also been an advisor to the Vietnamese, Bob was particularly interested in the book.

I also owe thanks to my agent and editors. Ethan Ellenberg sold the manuscript almost before I had cleared my desk. Jared Kieling and Jesse Cohen, of St. Martin's Press, were absolutely outstanding. After reading several books on the publishing business, I had expected tough and demanding editors. They were neither and graciously helped me through every step of the editing and production process. Thanks guys.

I may have made a mistake when I sent a copy of the first draft to my son, Bob, who was attending the Artillery Officer Basic Course at Fort Sill, Oklahoma. I kept getting enthusiastic reports as it passed from one lieutenant to another and I got a little concerned. I had to remind my son that he and the others were in a different army. If he tried to do some of the things his father had done, his army career probably would come to a screeching halt, terminated by a colonel just like me. To paraphrase General Douglas MacArthur: old soldiers don't die—but they do get a hell of a lot smarter.

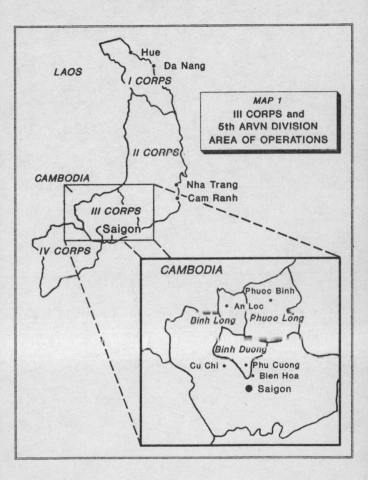

MAP 1
III CORPS and
5th ARVN DIVISION
AREA OF OPERATIONS

1 THE SAIGON RIVER

It was pitch black in Phu Van, but the trucks cast enough light for the companies to assemble and for the NCOs to check their men. There was surprisingly little confusion as the heavily loaded soldiers stumbled out of the darkness and joined their platoons. Although it was four o'clock in the morning and everyone was tired, the troops had done this so often that it was almost instinctive.

The rucksack straps began digging into my shoulders, and I realized that I had been a little too eager to get going. The other three Americans had been through this routine before, so their equipment was piled on the ground next to the jeep. Except for an unproductive walk in the woods the previous day, this was my first major operation, and with my new jungle fatigues, shiny boots, clean camouflage helmet cover, and pale-white skin, I knew that it looked like it. The black bar on my collar didn't do anything to offset my obvious lack of experience. It was 1967, and the army was promoting officers to first lieutenant after only one year in the service. Although I had just come from commanding a rifle company, that assignment had been in the States, and we'd only fired live ammunition at targets on the

1

range. The Ranger tab and parachute wings helped a little, but the only badge that counted in Vietnam was the Combat Infantryman Badge (CIB), and I wouldn't be eligible for that for thirty days. Unfortunately, I looked every bit like the *fanugee* [fucking new guy] I was.

The others standing next to the jeep didn't have the same problem. They'd been doing this for months and probably couldn't have cared less about how they looked. My new boss, Captain Chuck Kinsey, was an impressive, good-looking West Pointer, but he was not much of a conversationalist. In the two days I'd been with the team, he hadn't spoken more than four or five times, and there hadn't been any backslapping, "Welcome to the team!" greeting. He had said that he was happy to get me, but I knew he'd wait until after my first firefight to make any real judgment.

Sergeants Likens and Mosley had given me a warmer reception, but I soon found that they were motivated more by the prospect of having another man to carry one of the radios than of getting a new officer. They were impressed that I had been a "buck" sergeant, but like Kinsey, they probably felt that nothing counted as much as how I would act when the lead started flying.

Likens was the junior NCO on the team, but he had been in 'Nam a little longer than Mosley. Because he always looked exhausted, it was hard to guess how old he was, but he was probably about my age—twenty-seven. Mosley was older, maybe in his early thirties. A solidly built black man, Mosley was a "lifer," a soldier who had found what he thought was a better life than the one he'd have in the Deep South.

Likens was no lifer, and he didn't hesitate to express his dislike for the army. He seemed to have a permanent case of the ass, and I'd only seen him smile once in the past couple of days. For some reason he didn't like the captain, and he made no effort to conceal his feelings. It probably had something to do with Kinsey's never volunteering to carry the heavy PRC-25 radio.

Mosley didn't seem to care for our boss either, but he was much less blatant about showing his feelings. There hadn't been a lot to laugh about since I arrived, but once in a while Mosley would chuckle about something, and I knew that he was my kind of guy.

2

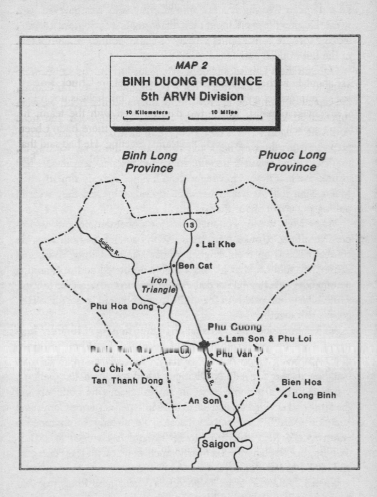

MAP 2
BINH DUONG PROVINCE
5th ARVN Division

10 Kilometers   10 Miles

Binh Long
Province

Phuoc Long
Province

(13)

• Lai Khe

• Ben Cat

Iron
Triangle

Phu Hoa Dong

Saigon R.

Phu Cuong

• Lam Son & Phu Loi

• Phu Van

Cu Chi •
Tan Thanh Dong •

Saigon R.

Bien Hoa
• Long Binh

• An Son

Saigon

I decided that I looked a little stupid wearing all my combat gear, so I dumped the rucksack and helmet in the back of the jeep. Although I'd only been wearing it a few minutes, my back was covered with sweat. Despite the excitement I felt about my first operation, I wasn't looking forward to humping seventy or eighty pounds of equipment in the hot tropical sun.

Once all the Vietnamese troops were loaded in the trucks, we were going to drive a few kilometers north on Highway 13 to Phu Cuong, the capital of Binh Duong Province. Shortly after first light, we were supposed to load into RAG (River Assault Group) boats and head down the Saigon River. After a couple of airstrikes and an artillery prep, we'd hit the west bank and sweep along the river, looking for the Viet Cong. It was a free-fire zone, and my new teammates had assured me that it was crawling with VC. Since nobody smiled when Major Man told us where we were headed, I figured they weren't pulling my chain about the bad guys.

Major Man was exceptionally tall for a Vietnamese and an impressive officer. Unlike Americans, ARVN soldiers sometimes wore combat ribbons on their fatigues. Major Man had three rows, and Kinsey had told me that he was a highly decorated and outstanding commander. Mosley had added that he had a good sense of humor, while Likens had paid him the ultimate compliment by saying, "He's got his shit together."

Man commanded the 3d Battalion of the 7th ARVN Infantry Regiment. (See the organizational chart on page 22.) When informed that I was being assigned as an assistant senior advisor to an infantry battalion in the 5th ARVN Division, I had screamed like hell. I'd expected to be a platoon leader in an American rifle company, but the army had decided to send me to the Military Assistance Command Vietnam (MACV), supposedly because I'd already commanded a company. I'd tried everything to get my orders changed to a U.S. division, but lieutenants don't carry much weight with the Pentagon, and all I'd gotten was a flat *no*.

Then I demanded to be assigned to a Vietnamese Ranger or Airborne battalion. Nobody objected to that, because those units had high casualty rates and always needed replacements. I was all set to buy a red or maroon beret, when some jerk in Saigon changed my

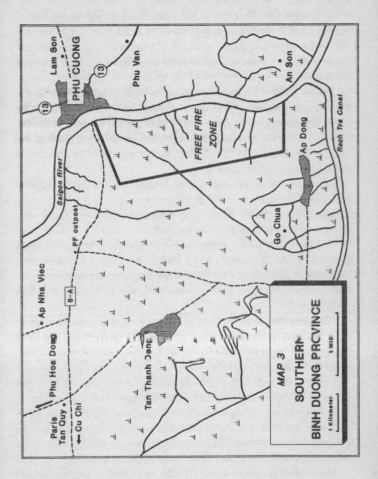

MAP 3

SOUTHERN
BINH DUONG PROVINCE

1 Kilometer
1 MILE

orders to the regular ARVN infantry. I protested, but the asshole refused to change the orders. When I got really obnoxious, he threatened to assign me to a desk job.

So I ended up with a tan infantry beret and an old World War II carbine, and I hitched a thirty-kilometer helicopter ride to the 5th ARVN Division headquarters, which was north of Saigon. Now I was about to find out if these Vietnamese draftees were as bad as the news media said they were.

The six week Vietnamese language course I had attended in the States taught me enough so that I understood Major Man when he yelled for the convoy to move out. I didn't relish the idea of driving Highway 13 in the middle of the night, but somebody at division had laid on the airstrikes for 0700. That meant we had to be loaded into the RAG boats shortly after dawn.

Despite my fears we pulled into downtown Phu Cuong without incident. Although it was still dark, there were quite a few Vietnamese up and around. Major Man's jeep stopped in front of a small cafe, and he motioned to us to go in for coffee.

The place was really dingy. The tables were covered with faded blue-and-white checkered plastic tablecloths. The walls were tiled with oriental designs and looked cheap. The concrete floor was dirty, and there seemed to be about an equal number of live and dead flies everywhere.

The proprietor scurried over to take our order. He was dressed in an undershirt, shorts, and sandals. Acting very honored to have Vietnamese and American officers in his humble establishment, he repeatedly bowed and smiled.

Shortly, he returned with our coffee, which was about equal amounts of espresso, canned milk, and sugar—and it was bitter as hell. As we began to drink, he showed up again, this time with a can of insecticide that he sprayed everywhere. (I noted that it was U.S. government issue, a small example of the thriving black market.) The mist settled on the surface of our coffee, adding more bitterness.

Major Man was joking again. "Today you kill some VC, Lieutenant Parrish."

I laughed but couldn't think of anything smart to say. I figured that as the new guy, it would be best to keep quiet.

6

After about twenty minutes, we got up and headed to the river. By this time the sun was just coming up, and more people were filling the streets. With all of our activity and preparations it was pretty obvious the VC knew we were going to conduct a river operation— I wondered if they knew exactly where.

When we arrived at the docks, I found a half-dozen LCM-8 landing craft and a couple of gunboats waiting for us. The gunboats, called "monitors," were about thirty feet long and built low to the water. They each had a 40mm gun mounted in a turret on the foredeck, a 20mm gun in the stern, and two .50 caliber machine guns mounted amidships. While I was impressed with their firepower, I wasn't impressed with their armor. It looked as if it was only about a quarter-inch thick, and that wouldn't stop much.

The troops were divided up and loaded into the landing craft. It was a tight fit, and we could have used a couple of more boats. As we pulled away from the shore, I banged my fist against the side of the craft. Just as I had thought—it was only thin steel. With the way we were packed in, the VC would get a very high body count if they hit us with almost anything. I wondered what my wife would think if she were told, "Your husband was lost when his ship sank."

In this area the Saigon River was about 100 meters wide. There were fields, fruit orchards, and hootches on the east bank, and a few sampans and small boats were tied up along the shore. The west bank was deserted, with nothing but coconut trees, thick grass, and brush. It didn't take an expert to figure out which side the VC owned, and our boats stayed close to the eastern shore as we moved down the river.

We'd been motoring along for fifteen or twenty minutes when a flight of two F-4 Phantom fighter-bombers roared by at low altitude. In a moment we heard loud explosions as they dropped their bombs. The troops cheered and climbed up onto the sides of the landing craft to watch. The aircraft made repeated passes, bombing and strafing the west shore about a kilometer downstream. Gray-black clouds boiled up from the target, and chunks of palm trees flew into the air.

This went on for ten or fifteen minutes, and then the planes were gone. Almost immediately I heard the boom of artillery, firing from Phu Cuong. A second later the 105mm rounds screeched overhead and exploded in the same place the fighters had bombed. I was a little disappointed, because the artillery rounds were not nearly as impressive as the bombs had been. Each time they hit, sprays of mud flew straight up, but the best part was the tearing sound they made as they passed overhead.

I had moved forward to the bow of our landing craft and was hanging on the ramp to watch the show. I thought, "Jesus, this is just like World War II in the Pacific." I was really enjoying myself when Likens pushed through the crowd and joined me.

"You know that when we go in, they'll drop the ramp on the bank. There are booby traps along the whole shoreline, and when the ramp comes down, it'll probably set one off. We had ten troops blown to hell the last time we did this, so I recommend we move as far back as possible."

So much for hitting the beach in the first wave. I followed him to the stern of the boat.

The artillery shifted its fire inland and our boats started moving. As we came abreast of the objective, everyone on that side of the boats opened fire. The noise was deafening as we poured thousands of rounds into the bank. There was no way any VC could have survived the bombs, artillery, and this barrage. Then I remembered that the same thing had been said about Guadalcanal and Tarawa.

On signal, the boats turned and headed toward the bank. The troops stopped firing, but the gunboats and artillery continued to shoot. We braced ourselves as the landing craft slammed into the bank. Then the boat crew slowly dropped the ramp, hoping that any booby trap would explode before the ramp opened completely.

No explosion. We moved forward, climbed onto the bank, and dropped over the other side. The bank was a dike about five or six feet higher than the abandoned fields beyond. Likens yelled at me to watch for booby traps, and I was careful to move in the footsteps of the troops in front of me. I heard an explosion off to my right. Somebody had hit a trip wire, and several soldiers were badly wounded. One had lost his leg.

A soldier near me grinned and pointed in the bushes. There was a hand grenade tied to a stick, with a trip wire disappearing into the grass. He said, "Beaucoup [many] VC." Boy, he sure knew how to get my morale up!

We moved straight in for about 100 meters and stopped. Apparently we had gotten through the main booby trap area, and it was time to organize for the sweep. Kinsey came up and said, "You and Likens move out with 9th Company; it will be in the lead. Battalion headquarters and the other two companies will bring up the rear."

I nodded and he moved off. Likens growled, "You better get used to it. You *always* get to go with the lead company. The captain always moves with the battalion command group, much further back. Mosley and I switch off between the two of you."

I laughed. "So I have the best chance of getting greased. You and Mosley have about half the risk that I do, and Kinsey has the least to worry about."

"Ya got it, *Loo-tenant*."

"Okay, now that I know how expendable I am, what do we do now?"

"We just wander along with the company command group and do whatever comes naturally. Oh, by the way, do you know how to adjust artillery?"

"Yeah, but I don't know if I can hit anything."

"It don't matter. I'll advise you while you advise the company commander."

"Thanks, I'll need that."

We walked off looking for 9th Company's commander, finding him just as the battalion was about to move out. He greeted me with a smile and a handshake. I said hello in my very rough Vietnamese, and we started out—no time for pleasantries.

This whole area had been abandoned for many years. Originally it had been planted in sugar cane and pineapples, but now it was growing wild. Every 200–300 meters there were tree-lined streams that ran into the river. Some of them were both wide and deep, and they made travel parallel to the river very difficult. About a kilometer inland the terrain turned into rice paddies. Likens told me that this was a traditional VC area and that we could usually count on a fight whenever we operated here.

9

After an hour or two of cautious moving, we stopped for a break. The sun was hot and we were drinking a lot of water. We spotted a hootch, and after checking it, we went in to get out of the sun. It wasn't a house, merely a palm-thatched roof held up by four poles. I thought that it didn't look very old, and so I asked the Vietnamese lieutenant why it was in such good shape.

"VC built it and they still use it."

Well, at least that confirmed my suspicions.

Just then one of the company NCOs came up with two pineapples he'd found growing nearby. Everyone perked up at the thought of some cool, juicy fruit. One of the Vietnamese whipped out a bayonet and began to carve. This was great! Although we were sitting in the middle of "Marlboro Country," we were having a picnic.

I had just finished one piece when suddenly two bullets smacked through the thatch about three feet over our heads. The Vietnamese and Likens grabbed the ground, but I just sat there. I figured that this would be a good opportunity to show everyone that I wasn't shaky. I pulled out my knife, cut another piece of pineapple, and offered it to the prone company commander. Then I cut some more and offered them to the others. Everyone, except Likens, sat up. Between bites I said, "VC shoot very badly," and the Vietnamese roared with laughter. Then I finished my piece and said, "Okay, so much for this crap, let's get the hell out of here!"

We all grabbed our equipment and hauled ass to some trees a few feet away. Everyone was giggling when we reached cover . . . except Likens.

"Goddamn it, L-T [lieutenant]! You're going to get your ass blown away doing that kind of stupid shit!"

"Yeah, I know, but I've got to show these people I'm not going to panic when the shit hits the fan."

"Whew! Ya gotta learn there's a fine line between being brave and being stupid."

"I thought they were basically the same thing."

Likens gave me one of the few grins I'd ever seen on his face. I hoped I'd made the right impression, and I hoped that word of my performance would be passed around the battalion—I didn't want to have to do that kind of dumb stuff with every company.

The company commander gave the word and we moved out again. The sniper had gotten everyone's attention and the troops were visibly more alert. After about a kilometer we reached a stream. This one was about four or five meters across and too deep to wade. The soldiers in the lead platoon stripped off their packs, and a couple of men swam across with a rope. When it was tied to a tree on the other side, the platoon began to cross. After the platoon reorganized on the far side, the company command group began crossing. I really wasn't looking forward to getting soaked, but I didn't see any alternative.

Likens called to me and said he'd found a fallen tree that we could cross on. The lead platoon had already begun to move across the next field, but I figured we could catch up to it. Although the tree was several feet in the air, it was possible to climb up and shinny across to the other side. Likens started off and I followed. It was slow moving, but at least we wouldn't have to swim.

Likens had just reached the far side and I was over the middle of the stream when all hell broke loose. Bullets were ripping through the branches all around me. We'd been ambushed and there I was, sitting on a log three feet in the air!

If I had had more experience, I would have dropped off into the water where the stream bank would give me cover. But I was new and stupid, so I continued to shinny across, covering the second half of the log in a fraction of the time it had taken me for the first half.

Chunks of wood and dirt were flying everywhere when I hit the ground on the far side. The field we were in had originally been planted in sugar cane, but now it was overgrown with cane leaves about chest high. It provided concealment if you were lying down, but it offered no cover. Likens was lying there with a look of fear on his face. I don't know what I looked like, but I wasn't afraid. This was just like a movie, and I couldn't comprehend that I might be killed. I knew that everyone else could be killed—but not me.

Likens was on the radio, reporting our situation to Kinsey. From what we had seen from atop the log, about thirty men in the lead platoon had just reached the middle of the field when the VC opened up from the far tree line, about 300 meters away. I couldn't tell how many VC there were, but Likens estimated about a company. I did

11

know that they were putting out a lot of fire and had several machine guns. I hoped that they didn't have mortars.

"I don't know how many casualties we've got," Likens was saying, "but I'm certain they wiped out the whole platoon. At the moment your Five [my radio call sign] and me are the closest living troops to Charlie."

I appreciated Likens' concern, but what he had said was not entirely accurate. We were out front, but the rest of the company (the command group and two platoons) was only a few meters behind us. The troops were not jumping out of the ditch to charge the VC, intelligently taking cover instead.

"Let's get back in the damn ditch. We're not going to accomplish a fucking thing laying here in the open."

Likens nodded, and we crawled over the bank to relative safety. Some of our troops were firing, but I didn't figure they were hitting much of anything. Both sides were in good positions, and the field between us had become a no man's land. It might be possible to crawl through the sugar cane, but if the cane leaves moved, they'd give away your position.

I don't know how long we lay there, but eventually artillery began to fall on the VC's stream line. Somebody at battalion had been busy. About the same time, Kinsey and Mosley appeared across the stream behind us. Although Mosley almost drowned trying to get across with his radio, they finally made it to our side.

I pointed out where the VC were and where our platoon had been when Charlie opened up. We couldn't see any bodies, but I was pretty certain that our men had all been killed. The Viet Cong had put too much fire into the field for anyone to have survived.

Major Man and his group joined us. Kinsey and Man were both on their radios, calling for support. Since only a direct hit by the artillery would do any good, we needed an airstrike. A few 500-pound bombs would change the picture dramatically, or at least I thought so.

Pretty soon a pair of armed helicopters showed up and began to make rocket runs on the VC. They were fairly accurate, but 2.75-inch rockets don't do a whole lot of damage. They did get the VC to duck.

12

Kinsey pointed out a clump of large trees to our left-front, saying that Major Man had ordered 9th Company's commander to move the remainder of his unit there while the gunships were keeping the VC busy.

"You and Likens go with them."

I nodded and Likens grumbled. I looked around and saw the Vietnamese begin to move, so I jumped to my feet and ran like hell. I was breathing so hard that I couldn't tell if Charlie was shooting at me or not. We made it, but the rest of the battalion stayed where it was. I wished that I knew what we were supposed to do next. A few minutes later a small O-1 Bird Dog began circling overhead. That meant that we probably had an airstrike on the way. Now we would stick it to them!

Sure enough, two A-1E Skyraiders showed up. The A-1E was an old propeller-driven airplane, but it could carry a lot of bombs and stay on station longer than a jet. Just having it fly around was a great morale booster. F-4 Phantoms zoomed in and zoomed out so fast that you never knew where they were—or if they would be back.

One of the airplanes went into a dive, dropped a bomb, and pulled up. Big explosion, tree limbs and mud flying everywhere—it was great! Under the cover of the airstrike, Major Man and another company moved up to us. I felt a lot better having more troops around me. Orders were given for some of the soldiers to crawl forward into the field. This didn't make the people happy since there wasn't any cover. The VC couldn't see them, but if they shook the sugar cane, they were going to draw fire and probably get themselves blown away. Kinsey said, "Bob, you go with them."

I didn't know what I had done to piss him off, but it seemed as if he was trying to get me killed. *What the hell was I going to do in the middle of no man's land?* Without saying a word I began carefully crawling through the cane. After a short distance I found my first dead man. He had been shot in the chest, and his face was an odd yellow color. His eyes were clouded and glazed over. I recognized the look from deer I had killed back home. I suppose I should have been shocked or revolted, but I only had a strangely detached, almost clinical feeling. Unless they were friends, dead men look just like dead animals.

A few feet away was another body. This guy had been hit in the head, and the back of his skull was a mass of blood, brains, hair, and gore. Nearby was a live soldier. He hadn't been with the lead platoon but was one of the troops ordered to push the perimeter out into the field. He was so scared that he was shaking.

I crawled up and gave him a big smile. He was surprised as hell to find an American officer in the middle of this field. He hadn't had a choice about moving forward but figured that I did. (He was wrong.) I tried out some of my schoolbook Vietnamese: "Hello, how are you? My name is Lieutenant Parrish. What is your name?"

He acted impressed, but I'm certain he thought my pronunciation was terrible. I gave him a cigarette but warned him not to light it here—the smoke might give his position away. Then I crawled on.

I found another body and then another live soldier. I went through the same routine as with the first man and got the same results—amazement and gratitude. By the time the airstrike was over, I had found several more bodies and had chatted with three or four more soldiers. As I was talking to the last one, the VC sprayed the field with a machine gun. (So much for airstrikes destroying everything!) They fired high, so either they weren't aiming or they didn't know that we were in the cane. I hoped that it was the latter.

When the artillery resumed firing, I retraced my way back to the command group. When I got there, I told Kinsey that everyone I had found from our lead platoon was dead and that I had returned because I was running out of cigarettes. "Do you want me to go back, or would you like me to lead a goddamn assault on the VC positions?"

He looked a little sheepish and told me to stay right there. I crawled back to where I had left my rucksack and took a break. Likens was lying nearby and he said, "Ya done good, Five."

"I still don't know what I was supposed to do. It seemed like a pretty dumb order."

He smiled and nodded his head in agreement. Then Kinsey waved for him to bring his radio over, and Mosley and he changed places.

"My radio battery is getting low, and I've got to change it," Mosley said. "What the hell were you supposed to be doing out in that field?"

"Beats the shit out of me."

After smoking a cigarette, I crawled over to where Kinsey and Man were hunched over their radios. As far as I could tell, we had what was left of 9th Company and all of 10th Company on this side of the stream. Eleventh Company was still on the other side. Despite the airstrikes and artillery, Charlie was still periodically firing at us. It looked as if he had a large force, but I couldn't judge its size.

"What're we planning to do?" I asked.

"Nothing for the moment. Major Man is trying to reorganize the battalion."

"Why don't we send 11th Company up the stream to the west? When they're out of range, they can cross the stream and get into the VC's canal. Then they can move back and hit Charlie on his flank. The rest of the battalion can cover them with fire. With a little fire and maneuver, we'd have the VC between a rock and a hard place."

"That's a little complicated for an ARVN battalion."

"Complicated? Shit! The only problem with this maneuver is control. They've got our stream line to guide on going out, and Charlie's stream line to use coming back. We can use machine gun tracers to mark the left limit of our fire, and they can use smoke or do something else to mark their advance. Everybody's got radios for coordination. Hell, I'll go with them to help."

Kinsey looked at his watch and then said, "I don't think they can do all that before it gets dark."

I shrugged my shoulders and shut up. For some reason nobody wanted to do anything but lie there and watch the artillery. I gathered that everybody was more willing to let the VC get away than risk losing more troops. I knew that everyone was demoralized, but a little leadership would go a long way to correct the situation.

Likens motioned to me, and I crawled over to him. "The two best company commanders are in the 9th and 10th, and they're on this side of the stream. The lieutenant in command of 11th Company is very weak. If he were told to maneuver, he'd drag his feet until it was too late to flank Charlie. I don't think Major Man can swap 10th and 11th Companies. Two companies trying to cross the exposed stream would draw a lot of VC fire."

"So here we sit, waiting for Charlie to withdraw. That's not very professional."

"Well, Charlie *might* withdraw—but he also might decide to attack."

"Attack? He'd be foolish to leave his ditch and get out in that open field."

"When it gets dark, he might crawl out into the sugar cane and play a little hand-to-hand with our troops. Our guys are so scared that they'll panic if the VC starts sneaking around among them. Charlie might decide to do exactly the same thing you recommended we do, only in reverse. It might get a lot more exciting after dark."

Christ! I hadn't thought that Charlie might infiltrate or flank *us*. I figured that we had *him* cornered; now I realized that *we* could be in big trouble! I decided to crawl back out into the field. There was nothing for me to do at the battalion command post, and I felt sorry for the little soldiers in the cane. I was excited, and I still *knew* that I'd never be killed. The only problem was now I couldn't blame Kinsey for sending me out there.

There was a small irrigation ditch that was about twelve inches wide and six or eight inches deep. It ran straight toward the VC's stream line, so any cover it gave was more psychological than real. I crawled over the legs of the first dead man, who was now stiff, and on to the live soldier I had originally found. He was still there but less afraid than he had been. He grinned, and I gave him a piece of gum. We talked for a while, and then I moved on. I repeated this process with the second and third man.

As we were whispering, I heard a Dust Off helicopter coming in. It was one of the old ones that the U.S. had given to the Vietnamese Air Force. The pilot made his approach to land south of our stream, and I guessed that it was a medevac chopper coming in to take out some of our dead and wounded.

As he disappeared from my view, the VC opened up. *Jesus, they were putting out a lot of fire!* There were still a hell of a lot of them alive and kicking. They were aiming at the chopper, so all the rounds were passing well over our heads. With that much fire, I was certain that they'd shoot it down, and I kept waiting for the explosion. Our troops along the stream returned the fire, and I lay there watching the bullets going both ways. I noticed that the VC tracers were green or pink, while ours were red. A few minutes later the chopper pulled pitch and took off. Charlie hadn't knocked it down, but I was sure that it had taken some hits.

It began to rain, so I started crawling back. The artillery had reduced its rate of fire and was sending in one round every thirty seconds or so. It didn't seem to be doing much damage to the trees along the VC's stream, and only occasionally did I see any branches or wood fly into the air. Most of the 105mm rounds just kicked up mud, meaning that they were exploding after they hit the soft ground. This meant that much of the shrapnel was being absorbed by the dirt or shooting straight up into the air. Proximity fuses probably would have been better because the rounds would explode in the air and spray shrapnel down on the VC, but since we were so close to Charlie, they would have caused us casualties too. So much for 105mm artillery—it sounded better than it really was.

After a while the rain stopped and the sun began going down. It looked as if it was going to be an interesting night. As it grew dark, the artillery began firing illumination rounds, which cast eerie shadows that danced over the battlefield, adding to the surrealistic atmosphere. Periodically our people or the VC would fire a few rounds, just to let everyone know that we were all still there. Occasionally there would be short periods of heavy firing, but they didn't last long and silence would return.

About 2100, a "Spooky" aircraft showed up. Spooky was an old C-47 cargo plane that had been converted into a gunship. It was armed with 7.62mm miniguns capable of firing thousands of rounds a minute. For the rest of the night, Spooky flew in slow circles, dropping parachute flares and spraying the VC's positions. The gun made a sound something like *Buzzzz*, and it put out a thick stream of tracers. I could see a solid line of red reaching from the aircraft to the ground, and I knew that for every tracer I could see, the gunner was firing four rounds of nontracer ammunition. That meant that there were five times as many rounds going into the stream line as we could see.

Kinsey divided the radio watch for the remainder of the night. One man would stay awake while the others tried to get some sleep in the mud. My shift was from 0100 to 0300; I had been told that this was the time the VC usually picked to do something. I paid only slight attention to the radio—but a lot to the field in front of us. I replayed the day's events in my mind and was satisfied that I had proved to everyone that I would be okay in a firefight. However, something kept nagging me. I went over it again and again but couldn't figure out what was wrong. Finally I had it—all of the films that I had seen of actual combat were in black and white, and they always had appropriate background mood music. Our battle had taken place in color and in bright sunshine. More important, there hadn't been any music. That was the difference between what I had expected and what I had gotten. I thought that it would sure be an improvement if the army sent up a psychological warfare plane to play martial music over every battlefield.

When I woke the next morning, everyone else was up already. Since people were standing, the VC must have pulled out during the night. I had been so tired that I hadn't heard our people moving around. Major Man had pushed his people out to the VC's stream, and they had secured the area. The dead were being brought back to our command-post area. There were about fifteen bodies already wrapped in ponchos, and more were being carried in. There had been no survivors from the lead platoon, and we expected that our final count would be about twenty-five dead. Surprisingly, there had been very few wounded. The VC had opened up when the bulk of the battalion was still behind the stream bank and its trees. Everyone who had crossed the stream, except for Likens and me, had been killed outright—or they were killed later when the VC poured fire into the cane field.

Likens was standing, looking at the bodies. "Ya know, if we had been just a little quicker getting across that stream, we'd be in a couple of those body bags."

I couldn't think of anything to say, so I just nodded. I saw Kinsey and Man beginning to move, so I grabbed my rucksack and started after them. When we got to the VC's stream, we found dried blood pools, bandages, bits of clothing, a couple of Ho Chi Minh sandals, and a lot of empty shell cases—but no bodies. The stream line was dotted with dug-in firing positions, many with overhead cover of logs and mud. It was obvious that the VC had prepared their positions long before we arrived. They could have known our plans, but even without any advance warning, they had had over four hours to prepare for us. With the airstrike and artillery, there could have been no doubt where we had landed. Then it was just a question of which way we would turn—north or south. I wondered if there was a similar VC position a couple of kilometers south of our landing site.

The trees were shot to pieces, particularly where Spooky's miniguns had torn them up, and there were artillery and bomb craters all over. I couldn't understand why there weren't any bodies.

Kinsey said, "They always recover their bodies so we can't tell how badly we've hurt them. It's frustrating, but that's the way it is."

Tenth Company's Lieutenant Tien came up and said something to Man. The major turned, pointed north, and said, "They went that way," reminding me of an old cowboy movie.

After we'd recovered all the bodies, we called for choppers to evac-uate them. When the last chopper lifted off, the battalion got ready to move. Major Man had sent a platoon to follow the VC's trail, and shortly it radioed back that the unit had reached the river. Man gave the order and we started off. Without waiting to be told, I motioned for Mosley, and we joined the lead company. This time, 11th Com-pany had been given the point since it was the only company that hadn't taken casualties the day before.

We'd gone only about 200 meters when one of our soldiers hit a booby trap, badly wounding two men. I thought, "Shit! That's not fair. We deserve some sort of break after what happened yesterday."

Again we called for a Dust Off, which took about thirty minutes to get to us. We loaded the wounded men and started moving again. I had learned a valuable lesson—when Charlie puts out booby traps or sets up ambushes, he usually does it along trails. I'd been told this many times before, but it hadn't really registered until now. From that point on, I stayed off trails unless I had *absolutely* no other choice.*

The threat of further booby traps slowed us to a crawl, and it was pretty obvious that we weren't going to catch up with the VC. When we got to the river, the platoon leader briefed Man. From the tracks

---

* I consider that staying off trails was one of the reasons why I survived almost two years in the bush. The main reason was pure luck.

we figured that the VC had dumped their bodies in the river and then had headed off to the northwest. Likens told me that this was a pretty common practice. VC bodies were often found floating down the river. If he could, Charlie would bury his dead, but if he was under pressure, he would dispose of them any way he could. The main thing was to conceal the number of casualties he had taken.

We turned to follow the VC trail. After about two hours, we reached an abandoned road that paralleled the river about a kilometer to the west. Charlie had blown every bridge along the road, which hadn't been used for years. Its only function now was as a boundary between the abandoned sugar cane and pineapple fields along the river and the worked rice paddies farther west. The area between the road and the Saigon River was a free-fire zone—meaning anyone spotted there was fair game.

We moved north along the road, and by late afternoon we reached a bridge that was still intact and guarded by a PF (Popular Force) outpost. Trucks were waiting to pick us up and take us back to Phu Van. Because of our heavy casualties, the regimental commander had decided to terminate the rest of the operation.

As we drove back in our jeep, Mosley said, "Well, Lieutenant Parrish has got his CIB now. He sure didn't have to wait long."

Everyone shook my hand, and Kinsey said, "You did a good job."

I thought that was pretty high praise from a guy who never said very much. I also thought that he hadn't wasted much time before he put my ass out on a limb.

2 ATTRITION

During the rest of May and the first half of June, we continued running battalion-sized search and destroy operations. We'd go out for two or three days and then come back to the 7th Regiment's compound at Phu Van for a day or so. Our overall area of coverage was about 200 square kilometers, ranging from a few kilometers north of Saigon to well above the infamous Iron Triangle. Its western edge was near Cu Chi, and its eastern boundary close to Bien Hoa. The Saigon River, flowing from north to south, split the AO (area of operations) down the middle. The river was significant for two reasons: first, because there were always a lot of VC near it; and second, because it was the boundary between the U.S. 25th and 1st Infantry Divisions. Wherever we went, our division had to coordinate with one of these two divisions. This may have caused problems for the various higher headquarters, but at our level, we thought it was great. Scattered throughout the area were American fire bases, all with artillery. When we made contact, Kinsey or I could call for fire support from two or three different artillery units. If they weren't firing for their people,

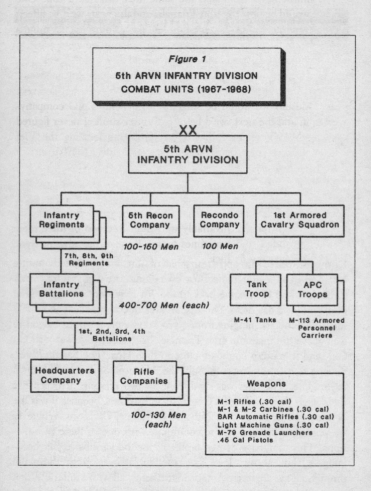

Figure 1

5th ARVN INFANTRY DIVISION COMBAT UNITS (1967-1968)

XX
5th ARVN
INFANTRY DIVISION

Infantry Regiments

5th Recon Company
100-150 Men

Recondo Company
100 Men

1st Armored Cavalry Squadron

7th, 8th, 9th Regiments

Infantry Battalions

400-700 Men (each)

Tank Troop
M-41 Tanks

APC Troops
M-113 Armored Personnel Carriers

1st, 2nd, 3rd, 4th Battalions

Headquarters Company

Rifle Companies
100-130 Men (each)

Weapons

M-1 Rifles (.30 cal)
M-1 & M-2 Carbines (.30 cal)
BAR Automatic Rifles (.30 cal)
Light Machine Guns (.30 cal)
M-79 Grenade Launchers
.45 Cal Pistols

they gladly blasted away for us. This was particularly handy because our own ARVN artillery was always short of ammunition.

We almost never went into the same area twice in a row. One time we would be near the Iron Triangle; and the next, we'd be fifteen kilometers south. Since the battalion had been doing this for quite a while, we could pretty much figure in advance if we were going to make contact and, if so, how big it would be. Some areas were crawling with main-force VC units, while others had only guerrillas. Division and regiment seemed to alternate us between hot and cold areas. One day we would be slugging it out with a VC company-sized unit; and the next, we'd be on a Sunday stroll. I never figured out why MACV claimed that it had problems locating the Viet Cong—all of us knew exactly where we could find a fight.

Had I not qualified for a CIB on my eleventh day in country, I could have gotten it six or seven times before my first thirty days in the field. Except for one of our rare successful night ambushes, most of our contact was *enemy initiated*—a euphemism for our being ambushed or surprised. Invariably we would lose two or three men killed or wounded, and sometimes more. None of the advisors were hit, but that was mostly luck.

Even when we were in a "cold" area, we had to be careful of booby traps. We lost a few troops to them, but our Vietnamese had a pretty good feel for where they were likely to be and were generally able to spot and disarm them before anyone was hurt. I learned pretty quickly where the VC typically put their booby traps (usually a VC-made hand grenade rigged with a trip wire), but I was much less skillful in locating their *punji* traps. For some reason, I was particularly unlucky and almost routinely fell or stepped into any punji trap in the area. Although I was never seriously hurt, my close calls were embarrassing. They happened so often that I began to refer to myself as "Punji Pit Parrish," figuring that if I poked fun at myself, others wouldn't.

The area along Highway 13 was heavily populated, with many small hamlets and villages stretching from Saigon to just north of Phu Cuong. VC main-force units generally avoided this area, moving through it only at night. Had they remained there any length of time, some villager would have reported their presence, and we or the Americans would have been after them. This did *not* mean that

23

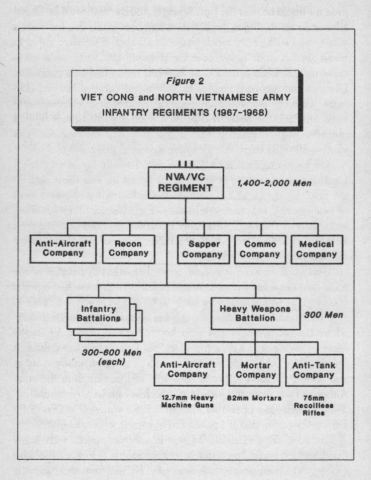

Figure 2

VIET CONG and NORTH VIETNAMESE ARMY INFANTRY REGIMENTS (1967-1968)

NVA/VC REGIMENT — *1,400-2,000 Men*

- Anti-Aircraft Company
- Recon Company
- Sapper Company
- Commo Company
- Medical Company

Infantry Battalions — *300-600 Men (each)*

Heavy Weapons Battalion — *300 Men*

- Anti-Aircraft Company — 12.7mm Heavy Machine Guns
- Mortar Company — 82mm Mortars
- Anti-Tank Company — 75mm Recoilless Rifles

Charlie left these villages alone. He had guerrillas working the area, trying to win over the people or intimidate them by assassinating officials, ambushing roads, collecting taxes, and generally making life uncomfortable. We swept through these areas a number of times but seldom made contact with anything more than a sniper or two. Infantry battalions do not make very good guerrilla hunters, and a five- or ten-man VC group wasn't about to stand and fight such a big force. These operations were intended more to show the flag than to add to our body count. I thought that we were wasting our time here, but at least I had an opportunity to learn more about the Vietnamese and their customs—particularly the one about not walking behind a hootch.

Farmers used human excrement to fertilize their fields, so their families saved the feces. Occasionally they would have a pot or bucket for this purpose, but usually they just answered the call of nature on the ground behind the hootch. At planting time, they would gather up the shit and spread it in their rice paddies. One quickly learned *not* to walk behind a hootch without being very careful. The cleated soles on our jungle boots were like magnets, and it was very difficult to clean the crap off them.

The area on the west side of the Saigon River was thinly populated, containing just a few small, widely scattered hamlets. Much of the area had been abandoned, and the fields were overgrown with thick foliage, providing a lot of places to hide. It was here that we could find the Viet Cong main force companies and battalions. In some areas, a twenty-five-man platoon wouldn't have lasted two days by itself.

In the northern part of our AO was the *pièce de résistance*—the Iron Triangle, the "Number Ten," baddest area of all. It was an inverted triangle formed by the Saigon and Thi Tinh Rivers, which joined about ten kilometers northwest of Phu Cuong. An abandoned east-west dirt road made up its northern boundary. The Triangle was about fifteen kilometers long by ten kilometers wide at its northern edge; it contained seventy-five square kilometers of thick jungles and bad people; and it was the largest free-fire zone in our area.

We took casualties almost every day. Sometimes we'd go for a

25

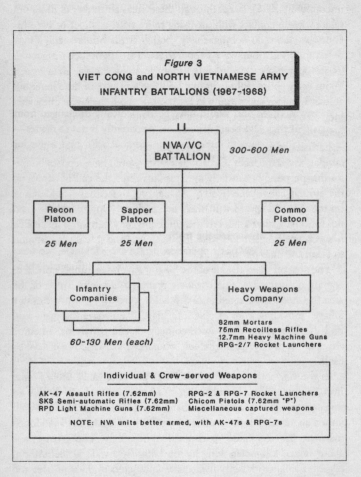

Figure 3
VIET CONG and NORTH VIETNAMESE ARMY
INFANTRY BATTALIONS (1967-1968)

NVA/VC BATTALION — *300-600 Men*

Recon Platoon — *25 Men*
Sapper Platoon — *25 Men*
Commo Platoon — *25 Men*

Infantry Companies — *60-130 Men (each)*

Heavy Weapons Company
82mm Mortars
75mm Recoilless Rifles
12.7mm Heavy Machine Guns
RPG-2/7 Rocket Launchers

Individual & Crew-served Weapons

AK-47 Assault Rifles (7.62mm)
SKS Semi-automatic Rifles (7.62mm)
RPD Light Machine Guns (7.62mm)

RPG-2 & RPG-7 Rocket Launchers
Chicom Pistols (7.62mm "P")
Miscellaneous captured weapons

NOTE: NVA units better armed, with AK-47s & RPG-7s

couple of days with no contact—and then all hell would break loose and five or ten men were hit. Besides affecting our morale, the casualties caused other problems. Typically, we would only get two replacements for every three casualties. Some replacements were new recruits, but others were soldiers who had previously been wounded and were judged to be well enough to return to the unit. The ARVN medical system had a very liberal policy on when a man could be returned to duty. Some of our returnees were so weak after months in the hospital that they couldn't hump with rucksacks and could barely keep up carrying only their rifles. I saw others who were permanently disabled and would have been medically discharged from the army if they had been Americans. Apparently it didn't matter— the Vietnamese doctors must have figured that we'd find a use for them. The practical effect of our casualty-replacement ratio was slow attrition. Every day the companies reported lower numbers of troops available for operations.

Since we seldom had any contact with other ARVN units, I assumed that all the division's battalions were having the same problem. Then one day I discovered the truth.

After a particularly nasty operation in the Iron Triangle, we were ordered back to Phu Van for a couple of days of badly needed rest. I had just had twelve hours of interrupted sleep when Kinsey woke me and said that Major Humphry, the regimental senior advisor, wanted to see me.

Although it was only midmorning, Humphry offered me a Vietnamese Ba-Moui-Ba ["33"] beer and motioned me to a chair. "Well, Bob, what do you think about your battalion? Your people have been having it a little rough lately, but they seem to be coping pretty well."

"It's a good unit, but at the rate we're losing people, we should be down to less than a couple of hundred effectives in another month or so. If we don't get some more replacements soon, we're not going to be worth a shit for anything but small patrols."

Frowning, Humphry said, "Yeah, I know. I'm trying to get something worked out with the regimental commander. Division has got to give you more replacements or change your operational assignments."

27

"What's the paddy strength look like in the other three battalions?"

"They're pretty much up to strength—probably around 85 percent of what they're authorized."

"You gotta to be shitting me! How come they've got so many people?"

"Ten of the twelve battalions are in static positions, mostly patrolling and sending out platoon-sized ambushes. Only your battalion and one up north are running search-and-destroy operations."

"If that's the case, why the hell are they at nearly full strength while we're out humping the bush with something around 50 *percent and going lower*?!"

"Well, as I said, I'm trying to get something worked out."

"Sir, you better do it quickly or division will have to disband us."

Humphry nodded and then changed the subject. "The reason I called for you is that I've got a little job for you. Captain Rogers, the 1st Battalion's senior advisor, is going on R&R next week, and I need somebody to fill in for him while he's gone."

"What about his lieutenant? Why can't he handle it?"

"He went home last week, and we don't have a replacement yet."

"Don't you have somebody else from one of the other two battalions? I hate to be away from the 3/7, particularly if they get in the shit while I'm gone."

"I'm afraid the other two lieutenants don't have enough experience to handle the job. The 1st Battalion's commander can be a handful when he wants to be, and he'd probably just totally ignore either of them."

"Okay, sir. When do I go and how long will I be gone? By the way, does Captain Kinsey know about this?"

"Yeah, I've already talked to him about it. I'll have a chopper take you over tomorrow. You should be back in ten days or so. Don't worry about your battalion; we're also going to use this time to give them a rest. Except for some local squad-sized patrols and ambushes, they're not going to be doing much."

"Okay, I'll get my equipment together."

As I started back to our hootch, Captain Robinson, Humphry's assistant, decided that he'd go along for a visit. As soon as we were out of earshot, he said, "Do you want to know why your battalion is

28

the only one that is always out beating the bush, and why you're getting so few replacements?"

Before I could answer, he went on. "Politics—there's a Colonel Quan at division that hates Major Man's guts."

"No shit? I've never heard anything like that. Man or his people have never said anything, and Kinsey has never even hinted that there might be bad blood."

"I'm not surprised. It's kept pretty hush-hush. Major Man is not likely to say anything, because if he did, it would force Colonel Quan to come out in the open and take some direct action against him. Quan is trying to get Man killed—or put him in a situation where Quan can get him relieved of command because his battalion took too many casualties in a fight."

"Christ! So Quan is willing to let a lot of our troops get killed or wounded just so he can screw Major Man?!"

"That's about the size of it."

I was stunned, but I didn't know what I could do. How could I complain through advisor channels? Major Humphry already knew about it and hadn't been able to do anything. Obviously I wouldn't get anywhere in the ARVN chain of command.

Later in the day, when Kinsey and I were alone, I asked him if he knew about Quan's vendetta.

"Yeah, I know about it. As far as I've been able to find out, it goes back several years, but I don't know how it started. I understand that Quan has repeatedly tried to get rid of Man, but so far General Thuan has refused to fire him. Humphry and I have talked to the division senior advisor, but we were told to stay out of it."

The next day Mosley drove me to Lam Son, and I caught a chopper to Tan Thanh Dong, where the 1st Battalion had its compound. Tan Thanh Dong was located about nine kilometers west of Phu Van, just beyond the free-fire zone along the Saigon River.

The 1st Battalion occupied a small but well-kept compound at the edge of the village. Its hootches were made of corrugated metal, with concrete floors—quite a bit better than what we had at Phu Van. It had the usual dirt berm, sandbagged bunkers, and barbed wire, but it was obvious that it had been a long time since the battalion had taken any VC fire. The soldiers that I saw seemed pretty carefree.

Many were walking in and out of the gate, wearing fatigues, soft caps, and sandals. Although they carried their rifles slung on their shoulders, it was clear that they weren't expecting any trouble. The difference between them and my beat-up troops was dramatic, and it made me resentful.

Captain Rogers met me at the helipad and took me to the team hootch to meet his two NCOs. Later he gave me a rundown on the situation. "We have the battalion headquarters and two companies here and a company-sized outpost on the southern edge of the village. During the day we usually have one company out patrolling. At night we send out a couple of platoons on ambush. Contact's pretty rare, so you shouldn't have any problem while I'm gone."

I gave him a hard look and pointing toward the east, said, "If your battalion wants contact, it only has to go a few kilometers in that direction. We've lost a lot of people over there in the past couple of months."

"There are them who fight and them who sit on their asses—this battalion is in the second group. There's not a damn thing you or I can do to change that. Believe me, I've tried."

Changing the subject, I asked, "What about Major Binh [Rogers' battalion commander]? What do you suggest I do? I don't want to sit here for two weeks with my thumb up my ass."

"Binh is a little tricky. He either likes someone or he doesn't. If he doesn't like you, he won't have anything to do with you. I don't know exactly what to tell you—you'll just have to play it by ear."

The next day, after Rogers had left, I wandered over to the battalion command post. Major Binh shook my hand and introduced me to a couple of his officers. He spoke pretty good English and was willing to shoot the shit. I figured he was sizing me up, although I was certain that Major Man had told him good things about me. We talked about Vietnam, the war, and what was happening in the province.

"Well, I think things are going to get better pretty soon," I said. "I've been told the U.S. will be sending a lot more troops to Vietnam in a couple of months."

"Oh? Where did you hear that?"

"My uncle told me."

"Who is your uncle?"

"General Westmoreland."

Binh's eyes widened and he said, "General Westmoreland is *your uncle*?"

"Yeah."

"Then why do you go on operations?"

"I'm a Regular Army infantry officer. If I spent my time in Vietnam behind a desk, it wouldn't matter who my uncle was—my career would go nowhere."

"How long have you been a first lieutenant?"

"Oh, only about four months."

"Well, since your uncle is General Westmoreland, you should be promoted to captain pretty soon."

"No, it doesn't work that way in the U.S. Army. I'll have to wait a couple of years, just like every other first lieutenant."

"Oh, I bet you get promoted early."

In the evening Binh came by the advisors' hootch and gave me a lengthy briefing on where the ambushes were placed and his plans for the next day. After he left, one of Rogers' NCOs said, "Jeez! That's the first time Binh has ever come over here to brief us. Normally Captain Rogers has to track down the S-3 [operations officer] to find out what's going on. You must have made a big hit with him when you talked to him this afternoon."

"I told him I was Westmoreland's nephew."

"Shit, sir. Westmoreland is your uncle?"

"Hell no, but it sure got his attention, didn't it? I don't think I'll have any problem with Major Binh while I'm here."

For the next few days I spent most of the time working on a suntan. It was boring, but at least I was getting rested up. Then one night while I was sleeping, there was a series of massive explosions—the VC were mortaring the compound! I rolled out of my bunk, grabbed my carbine and a radio, and charged outside to find a bunker. When I cleared the door, I found everything just as usual. The guards at the gate were smoking cigarettes, and other Vietnamese were swinging peacefully in their hammocks. We weren't being hit—the air force was putting in an "Arc Light" strike off to our west. Two or three giant B-52 bombers were dropping seventy thousand pounds of bombs *each*. An Arc Light is impressive as hell, but when you hear it inside a tin

31

hootch, the sound is awesome. I watched until it was finished and then carried my stuff back indoors. Although the two sergeants chuckled, I wasn't embarrassed. To survive in Vietnam, you had to react quickly to anything out of the ordinary—and an Arc Light could certainly be classified as extraordinary.

The remainder of my time with 1st Battalion was comparatively uneventful. A few days before I was to leave, one of their night ambushes did make contact. It was obvious from the results that the ARVN soldiers were just as surprised as the Viet Cong. An ARVN soldier was carried back to the compound in serious condition—he had multiple shrapnel wounds from an M-26 grenade. Since the VC normally used homemade or Chicom (Chinese Communist) grenades, it was possible that he had been the victim of a grenade that had been poorly thrown by one of his own people. A search of the site the next day failed to turn up any evidence of VC casualties.

Finally, after I had grown tired of the good life, Captain Rogers returned from R&R, and I went back to my battalion.

3 JOINT OPERATIONS

When I arrived back at Phu Van, I found everyone packing. The battalion had received orders to relocate to a fire base across the Saigon River, a place called Paris Tan Quy. Because of the odd name, I expected to see some old French fort, but it was just an ordinary fire base, with berm, bunkers, and barbed wire. No hootches. We were to live in sandbagged underground, or partially underground, bunkers.

Running between Phu Cuong and Cu Chi was a dirt road officially known as Provincial Highway 8-A, but it was *no* highway and everyone just called it "Eight Alfa." Paris Tan Quy was located almost midway between the two towns. About a kilometer to the north was another firebase, which was occupied by the American 2d Battalion, 14th Infantry, from the U.S. 25th Division.

It seemed that it was now "career enhancing" for U.S. commanders to work with ARVN units, so arrangements had been made for us to conduct joint operations with the 2/14 Infantry. When we arrived, Kinsey, Man, and I drove over to the unit's firebase to introduce ourselves and coordinate on local defense and future operations.

We were warmly greeted by the battalion commander, who introduced us to his staff and company commanders. Since we just *happened* to plan our arrival for lunch time, he invited us to eat, and I wolfed down my first real American meal in a long time. (We normally ate whatever our Vietnamese troops did.) Kinsey was polite, particularly since he was a captain and the U.S. battalion commander was a lieutenant colonel. Major Man was charming but became a little uncomfortable when he tried to figure out what was on his plate. After lunch we returned to the operations center to discuss specifics.

The battalion commander claimed to want joint operations, but he said that he was concerned about how the two battalions might go about it. "Normally, a two-battalion operation would be controlled by a higher commander at brigade or regiment. We don't have that here. You, sir [gesturing to Man], and I are on our own. I'm certain it would cause a great deal of confusion if either you or I tried to command both forces. So I believe the best thing to do is to attach one of your companies to my battalion. It could work with us for a week or so, and then we could swap it for another of your companies. After a month all your companies would have had an opportunity to work with us."

This pissed me off, but I kept my mouth shut. I suspected that he really didn't want to have anything to do with us but was being forced to work with the "little people"—the Vietnamese. He'd figured out a way to have "joint" operations without really having joint operations. What his proposal actually meant was that we would give up one of our three rifle companies, while he kept all four of his own. If we agreed, we'd be limited in what we could do on our own, with only two companies.

Unfortunately, I was not surprised when Major Man accepted the colonel's plan. He was under equal pressure to work with the Americans—and he had an ARVN colonel who was just looking for an excuse to get him fired. The American colonel concluded the meeting by saying that he would have his operations officer work up some plans and we could get together in a couple of days to go over them.

I waited until we got back to Paris Tan Quy and had dropped off Major Man before I said anything. I told Kinsey that I thought that the idea was bullshit. He agreed, but he reminded me that we really didn't have any choice. He pointed out that there were some good aspects. Our companies would get experience working with American units, and when a company was not committed to the 2/14, it could run local operations and bring its replacements up to speed. There would also be time for some badly needed maintenance on our weapons and equipment. Finally, we'd show the Americans that at least some ARVN soldiers knew what they were doing and were pretty good fighters. I still didn't like it, but I acquiesced.

When we got back, we found Likens and Mosley lounging in a couple of cheaply made lawn chairs at the entrance to our bunker. They wanted to know what had happened at the meeting and what our plans were. Kinsey went into the bunker, and I gave them a rundown. Likens reached over and slapped Mosley on the shoulder. "We've got it made for a while. We'll have to take turns humping with Lieutenant Parrish, but when we're not out with him, we can soak up some sun and suds. *You're* [grinning and pointing to me] going to be working your ass off for the next month or so, but you *like* that combat stuff, don't you."

Likens asked about the 2/14 firebase. As I began to tell him about the defenses and the six tubes of 105mm artillery, he shook his head. "I don't care about that crap. Do they have hot chow?"

I told him that they did and that we had eaten lunch with them —steak, ice cream, salad, and all the usual "goodies."

"Come on Mosley, let's go do some of our own coordination with those guys."

Later that evening, when one of the Vietnamese cooks advised us that dinner was ready, we thanked him but told him that we weren't feeling too well and wouldn't be eating. He replied, "Major Man also sick."

Major Man and the four advisors spent the rest of the night running to the slit-trench latrine. After a nearly constant diet of bland rice and vegetables, the steaks had done a job on our digestive systems.

We spent the next several days working on the firebase defenses and repairing our equipment. Man sent out local patrols and ambushes, but there was little contact. We found signs that the VC were checking us out, but apparently they weren't ready to do anything just yet.

Since we weren't doing much, Major Man decided to have a big dinner for the battalion officers and advisors. The cooks went to the market in Tan Thanh Dong and bought all the fixings. In midafternoon a runner came to escort us to the tent where the party was to be held. All the officers were standing next to a long table, wearing their best fatigues. After much handshaking, Major Man motioned to us to sit down. In front of me was a shallow dish that seemed to contain a mixture of chocolate pudding and peanuts. Since we had always had very basic Vietnamese fare, I had no idea what it was.

Man stood up and made an introductory speech welcoming everyone, and then we turned our attention to the meal. As I picked up my big flat-bottomed Vietnamese spoon, I had the feeling that I was being watched. I glanced around and noticed that everyone seemed to be looking at me out of the corners of their eyes. They were trying to pretend that they were busy preparing to eat, but several of them were giggling and surreptitiously jabbing each other in the ribs. I knew that I was the object of their interest, but I couldn't figure out why. Then I took a taste of the "pudding" and knew. I swallowed the stuff, and keeping a straight face, I took another spoonful. The giggling grew louder, and then with a big smile Lieutenant Phuc, the S-3, asked me if I liked the food. I replied that it was *interesting* and continued to eat. Finally everyone roared and Major Man asked, "Do you know what that is?"

With a completely straight face I replied, "Yes, Major. It's some sort of blood and peanuts, but I don't know what the spices are." Then I calmly took another bite. Everyone stopped laughing—I had spoiled their joke.

"You *know* it's *blood*?"

"Sure, what kind of blood is it?"

"It's duck blood. Most *covans* [advisors] don't like it, and some even get sick when they find out what they're eating."

"Well, you know Americans. Some of them are *dinka-dau* [crazy]."

Man looked at me and said, "Lieutenant Parrish, you are Number One!"

Everyone applauded and began to eat. The cooks had outdone themselves, preparing seven or eight courses. Each time we finished one, they'd bring in another, and naturally the advisors were expected to eat plenty of everything. Some of the courses were Vietnamese, and others were Chinese or a variation of Chinese. Except for the duck blood, the meal was really very good. I had already developed a taste for *nuoc mam*, so that posed no problem.*

After the banquet Likens asked me if I *really* liked the duck blood.

"Hell no, but I wasn't going to let them make me the butt of their joke."

"How did you know what you were eating?"

"Well, it reminded me of eating peanuts with a nose bleed."

Two days later Mosley and I joined 9th Company for our first joint operation. Because we were going to work for the Americans, Major Man informed me that I was to *command* the company. Lieutenant Chung, the company commander, would help me by passing my orders to the platoons. Chung spoke moderately good English, and that, coupled with my ever-improving Vietnamese, would ensure that the platoon leaders understood what I wanted.

We moved up the road to the 2/14's firebase and got ready for the choppers that would insert us into our AO. Man, Chung, Kinsey, and I had already gone over the plan with the Americans, so we knew what was supposed to happen. One American rifle company and 9th Company would be choppered to a stretch of 8-A at the edge of our old nemesis, the area next to the Saigon River. (See map on p. 5.) Once on the ground we'd sweep to the south with two companies on line. A couple of RAG boats would patrol the river to prevent the VC from crossing to the other side. Four companies from the ARVN 1st and 3d Battalions would establish blocking positions on the west to try to catch any VC escaping in that direction. When we had pushed far enough south, the RAG boats would move into the Rach Tra Canal to block the (hopefully) cornered VC.

On paper the plan looked good, but it had one big weakness. The Rach Tra Canal was the boundary between III Corps and the Saigon Capital Military District. If we took fire from the canal's southern bank, we'd have to go through a lengthy chain of command to use artillery, helicopter gunships, or airstrikes. It would take at least thirty

---

*Nuoc mam was a brown liquid made from rotting fish. It smelled terrible but didn't taste too bad.

minutes to get approval, and in that time the VC could do a lot of damage and then slip away. I pointed out that it would be better to insert us in the south and have us move to the north, but this plan was rejected, because it would take too many troops to block the open area in the north. The colonel felt that using the RAG boats to block the canal would be a better use of our limited forces.

As we waited for the choppers, the American GIs were giving us the once over. This was their first experience working with ARVN, and they were highly suspicious. They'd heard about ARVN units getting their asses kicked badly and didn't want to have to depend on any Vietnamese for their survival. Although I resented it, I understood what they must be feeling. A few minutes later a company of UH-1D slicks (troop-carrying helicopters) landed and we climbed on board. The ride was short, since we were only going about seven kilometers down the road. When the choppers touched down, we un-assed them and moved into the brush on both sides the road. Setting up local security, we waited for the helicopters to return with the 2/14's rifle company. Naturally the LZ [landing zone] was cold. Highway 8-A was well-traveled during the daytime, so we hadn't prepped the LZ with either artillery or air. Two Huey gunships circled low overhead, ready to fire just in case, and two more were orbiting a little higher as backup. I suspected that we were being landed here so that the operation wouldn't start off with a firefight. The American battalion commander didn't want to take too many chances until we had worked out some of the bugs of operating together. In about ten minutes the choppers returned with Alfa Company, which set down without any problems.

I could see that Alfa Company was deploying, so I told Chung to get our platoons ready. When we were all ready, the battalion commander, who was overhead in a Huey, radioed for us to begin to move out. Alfa was on the left, with its flank on the Saigon River, and 9th Company was on the right. We both had two platoons on line and one platoon trailing. Alfa's trailing platoon was next to the river, while ours followed my right-most platoon. From the air the force looked like a wide, box-shaped U.

The terrain was mostly waist-high grass, with scattered bushes and palm trees. About every kilometer, there was a stream running parallel to our front line, with trees and thick undergrowth along its banks. We'd been moving for about thirty minutes when I heard a muffled boom off to my left. I had Chung signal to our troops to halt while I took the radio handset to find out what Alfa had run into. Alfa's captain was reporting to the battalion commander that his flank platoon had hit a booby trap. He had four WIA [wounded in action], one of whom was serious, and was requesting a Dust Off. I relayed this information to Lieutenant Chung and told him to let the men rest in place until the Americans evacuated their casualties.

It took about thirty minutes to get the wounded men out and another few minutes for Alfa to reorganize; then we began to move. We were guiding on the Americans and adjusting our speed to match theirs. This time the movement was much slower than before, because Alfa was paying more attention to possible booby traps. Despite their caution, we heard another boom in about an hour—one KIA and one WIA. It was about 1000, so I told Chung to let the troops eat while the Americans sorted out their problem.

After another forty-five minutes the battalion commander radioed for us to move out. I responded that my people were eating breakfast and would need another twenty minutes before they'd be ready. There was a long silence, then he radioed that we should let him know when we were prepared to move.

Finally, to the relief of the "round eyes" (Americans), I called that we were ready, and in five minutes we started off again. At noon Alfa called and told us to hold up for lunch. Of course my people had already eaten, so we just relaxed for an hour. At 1300 we picked up our rucksacks and resumed moving. We'd begun the operation five hours earlier and hadn't moved more than one kilometer. At this rate it was going to be several days before we got to the Rach Tra—even if we made no contact. I knew that the battalion commander was going to be pissed when I asked for a halt to eat the afternoon meal, so I had told Chung to have the troops cook enough rice for both the morning and afternoon. I'd figured that the Americans would hit another booby trap, and that we could eat in the time it took them to evacuate their casualties. Sure enough, at 1500 there was another boom. This time Alfa had only one casualty—at least they were learning to spread out more. It wasn't smart to bunch up with booby

39

traps around. I told Chung to have the troops eat quickly, so hopefully we'd be done before the Americans were ready.

Unfortunately, they got the guy out faster than I had expected, and I had to radio that we weren't ready. The colonel asked what the hold-up was, and I had to admit that we were eating. This obviously didn't set well with him, but he held his temper—at least over the radio.

After another hour of moving, we were told to halt and set up an NDP (night defensive position). We moved to the left to join Alfa Company, where Captain Jefferies, their company commander, showed me how he had positioned his people.

"I've set up a 360-degree perimeter, and you can tie in on the west side."

"How come you're covering a full 360? That'll mean you'll have some of your men facing my troops."

"Well, I'm concerned about your ARVNs. If we get hit tonight, I have to make certain my rear is covered."

This angered me, but I concealed my feelings. Chuckling, I responded, "I've been arguing with my Vietnamese company commander for the past hour. He doesn't want to set up a joint perimeter with you. He's afraid *your* guys might break and leave *our* rear open. I've had a hell of a time convincing him that you won't."

Sputtering, the captain said, "What do you mean? We're *Americans*; we're not going to break!"

Just then the battalion commander's chopper landed. He and his operations officer got out and walked over to us. "Are you two getting set up okay?"

Before the captain could say anything, I said, "Well, sir, not exactly. It seems that Captain Jefferies is worried about my troops breaking and running if we get hit tonight."

The colonel gave Jefferies a cold look and said, "What the hell is this? You and the ARVN are supposed to be working together."

"Sir, I was just concerned about my people's security."

"Well, knock off that shit and start cooperating."

Having made my point, I said, "I understand Captain Jefferies' concern—he's never worked with Vietnamese before. But I'm with them all the time, and I *do* know them—and more important, I *trust*

40

them. On the average, your people have probably only been in the bush for four or five months. My people have been out here for *years*. Besides their combat experience, they know the people, the terrain, and the VC. Your guys would do well to watch how my troops work."

Changing the subject, I thought that I'd better explain why we had held up the movement a couple of times that day. "Despite what you might be thinking, my Vietnamese are not *always* eating. We have a fundamental difference here. Americans eat three meals a day, while the Vietnamese eat only two. In the morning they have a light snack of bread and coffee or tea. Then they eat their first meal around 1000 hours. They have their second meal in late afternoon. We could adjust that to your schedule without too much problem, but there's another difference. Your people have C-rations. They can open 'em and eat 'em in about thirty minutes. My people carry rice. We have to stop, make small fires, and cook the stuff for twenty minutes before we can eat it. That's at least thirty minutes just to get it ready. It takes us about twice as long as your people to get something to eat."

"Oh, I didn't know that. If we got your people C-rations, would they eat them?"

"In a heartbeat! If you could get us C-rations, I could work their schedule around yours."

"Okay, that's what we'll do. My mess officer probably will complain about the extra rations, but we'll fix that. I'll have C-rations brought in tomorrow morning."

He turned to his S-3, and we began to discuss the operation. Jefferies complained that the area along the river was pretty heavily booby-trapped, and besides slowing him down, it was causing a lot of casualties. When he was done, I said, "I'll take that flank if you want, but you're going to find that the area on the right is also thick with booby traps."

"If that's so, how come your people haven't tripped any?"

"I told you—my people have been doing this for a long time."

"Okay, you guys switch areas tomorrow."

The colonel went on talking about the plan and warned us to expect to be hit that night. I told him that I didn't think that would happen. Charlie might drop a couple of mortar rounds on us, but there probably wouldn't be any ground attack. Puzzled, he asked why

41

I thought that. I explained that I had worked this area quite a bit, and felt that I had a pretty good handle on the tactics that the VC used here. I told him that we had caught Charlie off-guard when we choppered in this morning. It would take him a day to figure out what we were going to do. "By now he knows that we're going to sweep south along the river—which, by the way, is what *everyone* does. Tonight he'll set up positions, probably (pointing to the map) along that stream line. When we get about here, he'll open up on us."

The colonel looked impressed and asked me what I would suggest. I told him that there wasn't much we could do since we were committed to the original plan. "We could air-assault one of the companies behind him and catch him between the two units, but that would only work if we *knew* that he was going to be along this stream line. If he was in the next one, we'd be splitting our forces and asking for trouble. About the only thing we can do is put an artillery prep on each stream before we move to it. We can shift the artillery south when we move in close. We'd need a couple gunships to be standing by to give us cover to get as near as possible. Then if we drop our rucksacks and charge the last bit, we might get into them before they could really hurt us. Are your guys up to a little 'hand-to-hand'?"

The colonel and his S-3 accepted my suggestions without hesitation. They should have—my plan wasn't very original. I had thought about a flanking movement with one of the companies, but at this stage the working relations were such that I was afraid that it would be too much to coordinate.

When the colonel left, I returned to our part of the perimeter where my people were digging in. On my way, a couple of American soldiers stopped me and asked if the ARVN "were really worth a shit." I told them to watch for the next couple of days and decide for themselves.

The night passed without incident, and early the next morning a chopper brought our C-rations. As I expected, the Vietnamese reacted as though it were Christmas. I explained the eating situation to Chung, and he told his platoon leaders to plan on eating around noon and 1700. With the C-rations, there was no objection.

About 0830 we began to move out, and by 1045 we were approaching the stream where I thought we would find our first VC.

The battalion commander told us to hold up while he got the artillery going. A few minutes later the 105s began firing from the firebase. It took them a few rounds to get on target, and then they started plastering the tree line. We dropped our rucksacks and moved slowly forward, crouching as much from the artillery as from any VC that might be in the stream. We got as close as we dared and then shifted the artillery, brought in the gunships, and started our assault.

At first we didn't take any fire, and I wondered if I'd picked a stream too short—or maybe the VC had di-di'd (ran) out of the area. But then a few VC opened up with sporadic and poorly aimed fire just before we hit the tree line. Almost 200 of us blasted away as we charged. We got to the stream and began cleaning out the few bunkers that still had VC in them. The colonel radioed from his chopper that a bunch of VC were running through the tall grass to the south, so we sprayed the field with fire. Since our artillery was falling along the next tree line, I radioed back that whatever VC we didn't hit would soon be turning to the west. If possible, the gunships should try to get them. As it turned out, this wasn't possible. The choppers would have had to fly through the artillery's gun-target line, and they might have been shot out of the sky by one of our own 105mm rounds. If we stopped the artillery, the surviving VC would just continue south—so some of the bastards got away.

Even though we didn't get them all, we scored pretty well. Two of Alfa Company's troops were lightly wounded, and we had four casualties, one of whom was killed. We got ten VC bodies out of the stream line and found seven more in the field. There were blood trails too, indicating that we had at least wounded some more. These were good results and everyone was pretty happy.

The battalion commander landed his chopper and repeatedly shook my hand. "That was great! How'd you like a full-time job working for me?"

"Thanks sir, but even if they'd let me go, I'd hate to leave my Vietnamese."

"Hell, son—I want them, too!"

That evening the atmosphere in the NDP was considerably different from what it had been the previous day. I couldn't tell where my lines stopped and Alfa's started. The Vietnamese and American troops

43

were all together, comparing weapons and equipment. It was a kick watching them trying to communicate with each other. Neither side spoke the other's language, but everybody seemed to be doing pretty well with gestures, pantomime, and a lot of smiling and laughing. I told Jefferies that we would have to separate them before dark so we could maintain control, but he said, "Let them get to know each other. We're a team, aren't we?"

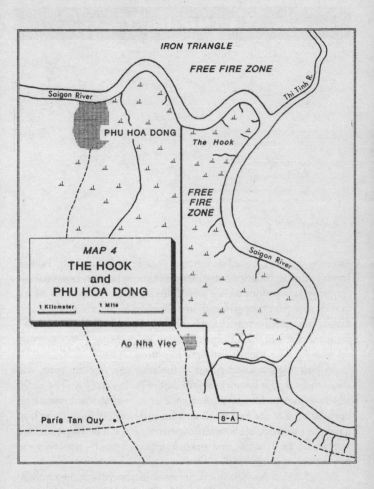

IRON TRIANGLE

FREE FIRE ZONE

Saigon River

Thi Tinh R.

PHU HOA DONG

The Hook

FREE FIRE ZONE

Saigon River

MAP 4
THE HOOK
and
PHU HOA DONG

1 Kilometer   1 Mile

Ap Nha Viec

París Tan Quy •

8-A

4 THE HOOK

After the choppers landed on the road outside Paris Tan Quy, I shook hands with Lieutenant Chung, congratulating him on how well his company had performed; we were both pleased. He took his troops to their sector on the perimeter, and I walked to the advisors' bunker. Likens was packing his rucksack when I entered. "I hope ya didn't get worn out with 9th Company, 'cause we're taking 10th Company up near Ap Nha Viec at first light tomorrow."

Ap Nha Viec, a small village at the edge of a free-fire zone, was about three or four kilometers northeast of Paris Tan Quy. The zone, which was about four kilometers wide at its widest point, ran along the west bank of the Saigon River from just north of 8–A to the Iron Triangle; a distance of about ten kilometers. It was a *bad* area, ranking just behind the Triangle, and we could expect contact—possibly some heavy shit.

We all hated Ap Nha Viec. In most of the villages in southern Binh Duong Province, the people were friendly. Whenever we moved through one of them, the children would run along beside us like a pack of dogs, inspecting the big hairy American advisors and begging for candy. If you smiled and nodded to one of the adults, he'd usually smile and nod back. The older girls would giggle and divert their eyes. Some of our soldiers had relatives in these villages.

Not in Ap Nha Viec. The people were sullen, and their kids never came out to get candy if we offered it. Even the very little children just stood there and stared at us with expressionless faces. These people were the families of the Viet Cong who operated out of the nearby free-fire zone.

"Yeah, I know," I said in reply to Likens comment. "I hope 10th Company does as well as 9th Company did. They were really professional."

I dumped my rucksack and began cleaning the M-16 rifle that I had finally managed to scrounge. There was one good thing about working with the 2/14—I could get ammunition easily. When we were by ourselves, I would have to carry extra ammo because there was no 5.56mm in our supply system.

The next morning we were off again, this time with the Americans' Bravo Company. No choppers this time—the operation began as soon as we crossed the road. There was much less tension than there had been when we had started with Alfa Company. Apparently the troops in Bravo Company had either talked to their buddies in Alfa or had heard reports on us.

Because the terrain was much less open, we moved in two columns. We stayed close enough together to coordinate movement and support each other, but far enough apart to allow some freedom for maneuver. It was pretty easy going, and by noon, when we stopped for chow, we were only about 500 meters short of the north-south road leading to Ap Nha Viec. I talked by radio to both the colonel and the Bravo Company commander. They were pleased about our progress—particularly by the fact that we hadn't run into any booby traps. I warned them that when we crossed the road, we'd really be in Marlboro Country, and that this area was as bad as the one we'd been in with Alfa Company.

We'd gone about 300 meters when I spotted a piece of paper hanging on a bush. Our lead platoon had seen it but had moved on past it. It was a page from a small notebook, enclosed in a plastic bag to keep it from dissolving in the rain. Written in Vietnamese was the warning: "Beware of this trail because it has booby traps. Do not walk down the trail because it is dangerous." Some Viet Cong didn't want his family or friends in Ap Nha Viec to get blown away. After we'd moved a few hundred meters farther on, the colonel radioed for us to halt and set up for the night. It was still early, but we needed the rest. The two companies moved toward each other, and the company commander and I laid out the perimeter for the platoons.

I'd learned that it was dumb just to stop for the night without first checking the area, so I asked Lieutenant Tien for a squad to do a little reconnaissance on my own. Tien wasn't happy that I was going out scouting, and he reminded me that it wasn't a job for an advisor. "No sweat," I assured him. "I won't be gone long." Tien shook his head, but I got my squad.

After about forty-five minutes of finding nothing, we started back to the NDP. I didn't know why, but I had an uneasy feeling. I took about ten steps and felt it—a wire! It tightened across my leg, just below the knee, and then released. Instantly, I realized that I had just tripped a booby trap. I lunged forward and threw myself on the ground, hoping that I could escape most of the fragments, but I knew I was in big trouble.

I lay there waiting for the thing to explode. When it hadn't gone off immediately, I knew that it must be a regular hand grenade with a four second delay—*Damn it, I should have run.* I might have gotten out of range in four seconds, but it was too late now. I kept lying there, but nothing happened. Finally I pulled my arms off the back of my head and looked at my watch. Its second hand was moving as slow as hell, but eventually it traveled a full ten ticks. Still no explosion, so I got up and ran like hell!

Once safely away, I stopped and slumped to the ground. I was scared and out of breath—but *very happy*.

I waited ten minutes and then carefully made my way back to the booby trap. Since it hadn't gone off after all that time, it probably would never go off. It was a cast iron, homemade hand grenade, serrated like our old ones. It was tied to a small bush, and the now-limp trip wire was lying near it. The safety pin had been pulled, but the spoon handle hadn't sprung completely free, so the firing pin hadn't travelled the full distance needed to strike the primer inside the grenade. I closed my hand around the body, squeezing the handle. I had one of the soldiers remove the pin from the wire and hand it to me. Then I inserted the pin in the hole, and the grenade was safe again. After cutting it free from the bush, I unscrewed the fuse and carefully pulled off the blasting cap.

When we got back to the NDP, Lieutenant Tien met us. I told him that we hadn't found anything worthwhile, but when I was finished, my squad began talking excitedly. After listening to their story, Tien turned to me and said, "You are a very lucky man, Lieutenant Parrish."

I took the grenade to where Tien had set up our command post and began inspecting it to find out why it hadn't gone off. The Viet Cong had used creosote to waterproof it, and the tarlike substance had softened in the hot sun and run down the sides of the firing pin. When I tripped it, the firing pin spring hadn't been strong enough to overcome the friction caused by the creosote. Lieutenant Tien was right: I had been a very lucky man.

The night was quiet, punctuated only by a short but heavy rain. During my turn on radio watch, I lay there watching flares drifting on the horizon and listening to the familiar sounds of war: the rising and falling sound of choppers flying somewhere to our south, and the periodic impact of artillery rounds. Occasionally there were softer booms and explosions that could have been anything. I looked at my watch, and its luminous hands showed that I had about twenty more minutes before I woke Likens for his shift.

Bang! Somebody on Bravo's side of the perimeter had fired a claymore mine. There was a moment of silence, and then several M-16s opened up on automatic, and someone threw a couple of grenades. The firing continued for a few seconds and then died away.

Likens had rolled off his poncho and was holding his rifle, ready for whatever might happen next. After four or five minutes we both concluded that whatever it was—shadows, imagination, or Viet Cong—it wasn't an attack. I hadn't thought that it was because of the timing. Charlie usually hit between 2200 and 0200. He would need the period of darkness between nightfall and 2200 to organize and move into position, and then the couple of hours before dawn to withdraw and hide before it got light. That left him about six hours to get his job done. Since the claymore had gone off just before 0400, it wasn't likely that it was a major attack, because there wouldn't have been enough time to do us a "damn-damn" and get away before the aircraft could get up to look for him. Units were hit at this time of night, but seldom in force.

When it was light, Bravo sent out a squad to check the area where the claymore had been fired. The men had only been gone a couple of minutes when they started hollering—it had been the VC. I picked up my rifle and wandered over to where they were laughing and scratching. The foliage in front of where the claymore had been placed was shredded, and so was the black-clad body a few meters away. He'd taken the full force of the explosion, and he was a bloody mass of ground meat and guts. He was wearing a canvas ammunition vest containing several mangled AK-47 magazines and a VC pistol belt.

The soldier who'd fired the claymore was both frightened and happy. He was scared because he realized that Charlie had been crawling up on *him* during the night, but he was happy because his buddies had been giving him a hard time about being spooked by shadows.

They searched around the body but found no weapon. Since the VC had AK-47 magazines, the dead man must have been carrying an assault rifle. It wasn't there, so it meant that he hadn't been alone. His unhurt comrades had waited until the firing stopped and then retrieved the weapon. We were going to have to be extra careful today because Charlie knew exactly where we were.

Despite our worries, we humped for the next two days without any contact. It looked like Charlie was withdrawing north to into an area that we called the "Hook." This wasn't the well-known "Fish Hook" along the Cambodian border; it was our own local version. This Hook was formed by a large sharp bend in the Saigon River east of Phu Hoa Dong, and it was just across the river from the Iron Triangle.

On the afternoon of the second day, a chopper dropped the colonel into our NDP to discuss a change in plans. Everyone had concluded that our operation had probably pushed the VC into the Hook, so it was time to come up with something new. We were to continue moving north along the river. When we got to the eastern edge of the Hook, an additional 2/14 company would chopper into an LZ that we would secure. Then the three companies would move directly west across the Hook, toward Phu Hoa Dong. A Vietnamese Regional Force (RF) company would set up blocking positions along the eastern edge of the village, several kilometers from our starting point. The Saigon River formed a natural barrier on the north, and the southern edge of our sweep area would be screened from the air by slicks; gunships would remain on call.

I had several reservations about the plan. First, the area we were to sweep across was flooded rice paddies. There would be damned little cover for over two kilometers, and it would be tough going through the knee- and waist-deep mud and water. Second, trying to block the southern edge with helicopters would be difficult, particularly since the choppers would be flying through the artillery gun-target line. If we needed artillery, the choppers would have to move out of the area they were supposed to screen. The colonel agreed, but there was little else we could do. We couldn't do anything about the terrain, and there were no troops available to block in the south.

We had been having trouble with our radio because moisture had gotten into either the handset or the radio itself. I asked the colonel if I could send Likens back to Paris Tan Quy to pick up another one. Bravo Company was being resupplied by chopper, so Likens could catch a ride back to the firebase and then return.

Lieutenant Tien and I walked back to our part of the perimeter and went over the plan. It was pretty straightforward, and there really wasn't much we needed to do to prepare to execute it. After we were finished, I got out some C-rations and ate an early dinner. Then I leaned back on my rucksack to catch a little rest.

I don't know how long I had been sleeping when someone called my name. There stood Mosley with his radio, rucksack, and carbine. "Likens' chopper has been shot down. They've got gunships over it now, and they're bringing in a reaction platoon from 2/14 to secure it and get the survivors—or bodies—out."

"Jesus, when did that happen?"

"Just a little while ago. When we got the word, Captain Kinsey had me grab a chopper out here so you'd have a radio—and someone to keep an eye on ya."

"Any word on Likens and the crew?"

"No, but I saw the chopper on the way out here, and it looked to be in pretty good shape."

I got on the radio and called Kinsey, but he didn't have any more information. Since Likens probably wouldn't be back even if he was okay, I briefed Mosley on the plans for the next day. I had just finished when Kinsey called and said that they had gotten the crew out. Likens was bruised but didn't need medical attention.

The next morning we moved about a kilometer north and secured the LZ that Alfa was going to use. We had just gotten our security out when they began landing, four choppers at a time. Once they were all on the ground, the three companies began to move in column, parallel to the river.

Soon the firm ground gave way to abandoned watery rice paddies. It was slow going, and it was difficult to keep the troops from bunching up and creating good targets. Finally we got to our turn point and began moving west, with three companies on line.

Suddenly Charlie opened up with a heavy volume of fire, and we hit the mud. The VC had timed it poorly, because we were just short of a paddy dike when they began firing, allowing us to take cover behind it. The dike rose about eighteen inches above the water and only provided cover if we lay—or more accurately, floated—flat in the paddy. We were completely pinned down and had no idea where the VC were located or how many people they had. The colonel's chopper orbited overhead but stayed high enough to be fairly safe from rifle fire. Unfortunately, at that altitude he couldn't spot the VC's position.

Charlie was smart; he fired just enough to keep us down but not enough to give his position away. So we lay there, becoming more and more frustrated. The other company commanders and I were all giving the colonel opinions of where the VC might be, but each of us had a different idea—meaning nobody really knew. The colonel had called up a couple of gunships, and they were ready if we could just pinpoint the target. The colonel's bird flew low a couple of times, trying to draw fire, but the VC weren't having any of it. It was a standoff.

The longer I lay there, the more concerned I became. If we didn't break this impasse, something was going to happen, and it probably would be bad. The VC could start mortaring us. There might not be any mortars with this particular group, but it could bring some up pretty quickly. Charlie might decide to withdraw to new positions after dark and do the same thing to us tomorrow. If we stayed here, additional VC might cross the river from the Iron Triangle at night and hit us on our flank—then we could be in deep shit.

I radioed the colonel to be ready to watch for the VC to open fire. Then I jumped up on the dike, waved my arms, and yelled, "*Ho Chi Minh du ma!*"—which loosely translates as "Ho Chi Minh is a motherfucker!"

There was a moment of silence, and then the VC opened up with everything they had. Boy, were they pissed off! I jumped back and flattened out behind the dike. The troops on either side of me just stared. One said, "*Trung-uy beaucoup dinka-dau!*" [Lieutenant, you're crazy!] The others just shook their heads. Then I began to wonder if I hadn't committed a big *faux pas*. Very few South Vietnamese were willing to believe that "Uncle Ho" was responsible for the war. Most regarded him as a national hero, something on the order of our George Washington. They couldn't bring themselves to accept the fact that besides being a nationalist, Ho Chi Minh was also a communist. I had heard a lot of strong language about various North Vietnamese, but no one ever said anything bad about Ho Chi Minh. I decided I could worry about it later.

Unfortunately, my timing had been off. The colonel's chopper had been flying away from the VC's position when they fired. They stopped as soon as it began to swing back toward them, and I hadn't stayed on the dike long enough to spot them myself. So other than possibly offending my Vietnamese, I hadn't really accomplished a damned thing.

We continued to lie in the mud, frustrated. I heard some mumbling to my left, and then one of my platoon leaders jumped on the dike just as I had. He only got the words "Ho Chi Minh . . ." out of his mouth when a burst of VC fire almost cut him in half. He was knocked flying and hit the water with a splash. I crawled over to where he was lying on his back. Holding his head out of the water, I tried to figure out what I could do for him; he was in bad shape. I yelled to Mosley to see if we could get a medevac. He low-crawled over and gave me the radio handset. I called the colonel and told him about my wounded lieutenant. He said that he could get a Dust Off, but it wouldn't be able to get in to us—the VC would knock it down. I told him that I'd get the man to the other side of the tree line, about a hundred meters to our rear. Then the chopper could come in low from the east and be masked by the trees.

I stripped off the lieutenant's pack and web gear, and began to drag him through the mud. The dike provided cover while I was in the paddy, but when I got to the edge, I had to cross fifty meters of dry ground sloping above the paddy. I pulled the man over my back and grabbed hold of his arms. Then after taking a couple of deep breaths, I got to my feet and ran like hell. I could see the VC's bullets hitting the ground around me and the trees just ahead. Fortunately, the lieutenant couldn't have weighed more than about a hundred pounds, and the adrenaline was really pumping through my body. I probably didn't set any track records, but I made it into the trees a hell of a lot faster than I would have ever believed—and I hadn't been hit!

The medevac got there pretty quickly, and I loaded the man on board. He was still breathing, but I didn't think he was going to make it. After the chopper pulled pitch, I went back into the trees and watched the gunships rolling in on the VC. The colonel had been able to spot the VC's position when they had opened up on me. I sat behind a tree until I caught my breath, and then I ran and waded back to the dike where I had left my weapon and equipment. The gunships were making nearly constant rocket runs on the Viet Cong, and no one shot at me on my return trip.

After about fifteen or twenty minutes of gunship attacks, the word was passed, and we got up and started forward. We got as far as the next dike when Charlie recovered, and we hit the mud again. The 2.75-inch rockets may have shaken him up, but they hadn't taken the fight out of him. The gunships left to rearm and refuel, and the artillery began firing again. I looked up and saw a gray O-1 Bird Dog circling overhead. Gray Bird Dogs meant airstrikes, because the air force painted their FAC [Forward Air Controller] aircraft that color, while the army's were painted olive drab. A few minutes later the artillery lifted its fire and an F-4 Phantom screeched low in front of us. I watched as it released a high-drag bomb. This type of bomb had a large fin assembly that snapped open after it was released. The bomb left the aircraft at the same speed as the plane; then the fins popped out, braking its velocity. It almost looked as though the bomb stopped in midair, then glided slowly forward onto the target. High-drag bombs allowed low-flying aircraft to get out of range before they exploded.

The bombs exploded impressively, but I'd been through this enough times to know that they almost never killed all the VC. Most of our ordnance (bombs, artillery rounds, rockets, grenades, etc.) lost a lot of its effectiveness in rice paddies. Most of it went too deep before exploding, so the mud and water absorbed much of the blast. Napalm was always great to watch, but it had accuracy problems. It was not unusual to watch a napalm canister float beyond the target before it detonated.

When the last bombs were dropped, the F-4s began strafing with their 20mm guns. This was the least effective weapon they carried, but we moved forward, hoping the F-4s could keep the VC pinned down until we closed on them. We only moved another hundred meters before Charlie began firing again. *Goddamn it! These suckers were really hard-core!* We estimated that we had run into a full company. There were fewer of them now, but there were still enough to give us a lot of casualties, particularly since we had to advance across flat, open paddies in waist-deep mud and water.

Down we went and the cycle began again: artillery, gunships, airstrikes. By nightfall, having moved only about 300 meters, we were ordered to hold in place until the next morning. It was a thoroughly miserable night. We had been in the water so long that just getting dry was now a top priority. I suggested sending one company out to flank Charlie. If our men were very careful, they could move on top of the dikes without making too much noise. But for a variety of reasons, this plan was vetoed. We'd have to stop the artillery illumination, and the VC might use the darkness to their advantage. The dikes were wet and slippery, and it wouldn't be possible to move quietly. The VC had probably moved from their original position, and we would be splitting our force. All this was true, but the *real* reason was that many American GIs were not trained well enough to operate effectively at night in large units. A thirty-man platoon could do pretty well, but anything larger usually turned into a "goat fuck."

Little happened during the night—except for a couple of RPG (anti-tank rockets) that scared the hell out of us before flying harmlessly overhead. The next morning, without having had any sleep or food, we tried again. Just as we expected, the VC had pulled out during the night, leaving only a couple of stay-behinds to harass us. We neutralized those guys with artillery and gunships. About noon we emerged from the paddies and started up to higher ground.

At the edge of Phu Hoa Dong, we came up to the RF positions. The RF had built poncho hootches and were relaxing in their hammocks. This pissed everyone off, especially my Vietnamese troops. After our past two days, we didn't appreciate finding our "blocking force" picnicking in the sunshine. A couple of my troops were so angry that they kicked and punched some of the RF, much to the delight of the Americans who would have been in trouble had they done the same thing.

The colonel landed and talked with his two company commanders and me. We hadn't found much when we had searched the VC's position. They'd gotten their bodies out after dark, and the rain had erased most of the blood. He said that he and the FAC had counted between twenty and thirty bodies from the air—so "we'd done good." Our own casualties were fewer than ten, all wounded. I went back to Lieutenant Tien and told him that choppers would be arriving in about an hour to take us back to Paris Tan Quy.

I noticed a change in the Vietnamese. They had always been friendly, but now individual soldiers were smiling and waving at me wherever I went. Sometimes they would join me and walk along holding my hand. This was a custom that few American soldiers understood or felt comfortable with. To the Vietnamese, showing affection to a member of the opposite sex in public was not considered proper behavior, but it was quite acceptable for two men to hold hands while they walked. There was nothing sexual about it; it was just what good friends did. My Vietnamese troops were showing that they considered me a buddy.

Major Man had always been very friendly and liked to laugh and make jokes, some of which were a little esoteric to us advisors. We'd politely laugh, but we usually missed the point or punch line completely. I had never been able to tell jokes, but I did like to make humorous remarks, and I was given to double entendres. As I came to know the Vietnamese better, I made little funnies more often. As time went by, I was able to tailor my jokes to their culture, and they actually began to see the humor.

Major Man may have been something of a joker, but he was also very proper. He always conducted official business with Captain Kinsey, and unless it couldn't be helped, he never went around him to discuss official matters directly with me. But since I was the advisor who probably would be where the critical action would take place, Man would usually motion for me to join them when he and Kinsey were going over something important.

On the day following my return from the Hook, Man came to the entrance of our bunker and asked Kinsey to come outside. A few moments later, Kinsey came back in and told me that Man wanted to talk to me. Because of the way Man worked, I thought that this was a little odd, but I picked up my rifle and went outside. Man had a very serious look on his face, so I didn't know what to expect. He told me that Lieutenant Tien and several officers had talked to him about what had happened during the operation. I felt a slight chill run down my back and began to apologize for having yelled "Ho Chi Minh du ma."

Before I could finish, Man said, "Fuck Ho Chi Minh! He is *the* Number Ten Communist. I want to tell you that everyone says you were very brave, and I am recommending you for the Cross of Gallantry."

"Major Man, I was very *dinka-dau* for doing that, and I was not brave, because I wasn't afraid of being killed. One must be afraid to be brave."

"Oh yes, you were crazy when you stood up and dared the VC to shoot at you, but you were also brave when you carried Lieutenant Tan across the open field while they shot at you."

"No, I wasn't brave, because I didn't believe the VC could kill me—I still don't think they can—but thank you anyway. Is Lieutenant Tan going to be okay?"

"He died."

He shook my hand, and I returned to the bunker. Kinsey and the two sergeants also shook my hand. Man had told the captain that he was recommending me for an award, and Kinsey had told the others. Likens growled good naturedly, "Ya keep doing that kind of shit, and you'll get your next medal posthumously—and we'll get a new lieutenant that we'll have to train all over again."

The only thing of any significance that happened during that week was a VC ambush of a Revolutionary Development (RD) propaganda truck. The government's pacification program included sending small teams into relatively secure villages to win the people's "hearts and minds." These teams helped the people with community projects, some limited medical assistance, and miscellaneous other actions designed to improve their existence and make them more receptive to the heavy doses of propaganda dispensed with the help. One of these teams was working in Tan Thanh Dong, and it would run a truck mounted with loudspeakers up and down the road. Besides government propaganda messages, the truck played Vietnamese music, always at an earsplitting volume. Each day the truck would drive a little closer to Paris Tan Quy. The battalion didn't know why the RD team was coming up our way, but everyone agreed that we could get along quite nicely without the racket. Finally one evening, a little later than a prudent man would be out, the shrieking music stopped with a bang—a *real* bang! Charlie had set off a command-detonated mine, blowing the truck, its loudspeakers, and the three RD people sky high. For the next couple of days, a joke passed around the battalion that maybe it wasn't the VC who had gotten rid of the noise, but whoever did it deserved our thanks.

Since we were back on our own now, we took trucks from Paris Tan Quy and headed south through Tan Thanh Dong to the southern edge of the province. We were going to try something a little different this time. The battalion set up a command post (CP) in a small village and kept one company in reserve. The other two companies were to conduct independent search operations to the north. If we had no major contact, the battalion would relocate its CP and reserve company north the next day, and we would repeat the process. I was given my choice, so I picked the company on the left, only because the ground wasn't as wet there.

Shortly before noon, 11th Company and I were approaching a place called Go Chua. In the center of a clearing was an old ornate pagoda, its multicolored paint having faded long ago. In sharp contrast, a modern, two-story building constructed of beige concrete block stood a few meters away, looking like a shoe box. Surrounding the two structures was a ten-foot-high concrete wall, with a wrought iron gate on the side we were approaching. I told Lieutenant Vinh, the company commander, that it looked completely out of place, sitting all alone several kilometers from the nearest village. He said that he thought it was probably a Buddhist monastery.

He told the troops to take a break and pulled the cord on a bell next to the gate. In a few moments a robed monk appeared and asked Vinh what he wanted. Vinh said something and the monk disappeared. After four or five minutes, he returned with an older monk and unlocked the gate. The monks escorted Vinh, the radio operator, Likens, and me through the courtyard to the main building. Standing at the door was an even older monk, who made a great show of shaking all our hands. He invited us into a large room where twenty or thirty people were sitting at a banquet table. About two-thirds were in saffron robes, and the others were dressed in dark pants and white shirts.

The old man, who was apparently the head monk, insisted that we join them at dinner, and we reluctantly accepted. Vinh sent his radio operator back to the troops to tell them what we were doing and to have them eat their midmorning meal. The three of us took off our rucksacks and laid them by the door, but we placed our rifles along the wall, just behind our chairs. The table was laid with a big meal, which surprised me since I had always pictured Buddhist monks with begging bowls.

I glanced at Vinh out of the corner of my eye and noted that he appeared a little uncomfortable. I was a little ill at ease, too—but our discomfort was nothing compared to Likens's. I had seen this guy in the middle of booby traps and in some hairy firefights, but I'd never seen him more tense than now. He was clearly out of his element. The other people also seemed nervous, particularly the younger ones in civilian clothes. I suspected that it was because of the two well-armed American soldiers in their midst. They probably thought that we were barbarians who might do something bizarre.

We ate for about thirty minutes in almost complete silence. No one made any real attempt at small talk. After we had finished, the head monk ushered everyone outside to a large altar in the courtyard.

As we stood in a semicircle, the old monk, holding a smoldering *josh* incense stick between his hands, began bowing and chanting a prayer. When he finished, he moved aside, and one of his followers repeated the ritual. Then the oldest guy in civilian clothes did the same thing. They went through all the men in civvies first and then the monks. I wondered what I was going to do if they wanted me to take a turn, but I wondered more about the nonmonks. Some of them didn't seem to have any idea of what they were doing, and even I could tell that they were stumbling over the words in the prayer. I tried to glance at Vinh, but I couldn't get a good look at him. Likens looked like he was going to have apoplexy.

Vinh took his turn and then handed his josh stick to me. I could have punched him out on the spot! Knowing absolutely none of the words of the prayer, I merely walked to the altar, and holding the sticks as the others had, I made several slow bows. I went over to Likens and whispered for him to do the same thing, which he did— as quickly as he could. Then the old monk gestured for everyone to go back inside. As we did, I sensed that there were fewer of us than there had been when we had eaten. Small cups were passed out, and a couple of monks went around filling them with tea. There was a lot more milling around the room than I thought was appropriate for whatever solemn occasion this was supposed to be. Likens was a little more relaxed, but he saw this as his opportunity to escape and quietly told me that he was going outside.

After a few more minutes of milling, Likens came back in and whispered that Kinsey was trying to contact me on the radio. I made my way to the door and over to the main gate where Likens was standing; he had moved there to be as far away from the party as possible. Taking the handset, I radioed Kinsey.

"Curly Tappet Six, this is Curly Tappet Five. Over."

"Five, this is Six. We've just received an intelligence report that the Victor Charlie province chief for Binh Duong is holding a meeting somewhere near Go Chua. Have you noticed anything unusual? Over."

"Six, this is Five. That's a Roger. The Victor Charlie province chief is having a big party."

"Say again your last transmission."

"I say again. The Victor Charlie province chief has been having a big party at the pagoda . . . and we've been his guests."

Now I knew what had been wrong! It was those guys in their best civilian clothes—they were nervous, and just about as lost with the Buddhist ritual as I had been. I grabbed Vinh's second-in-command and told him to quickly surround the pagoda and stop anyone trying to leave. I told him that there were VC inside, so he should instruct the troops to be prepared to shoot.

Acting nonchalantly, I went back inside to get Vinh and found him talking to several monks. I glanced around the room, looking for the VC, but there were none left. They'd all slipped away. I told Vinh that he was wanted outside. When we were away from the building, I told him what Kinsey had said and that I had ordered his troops to surround the compound. At first he gave me a blank look, and then he realized what had been happening.

We were much less polite when we searched the pagoda than we'd been when we were honored guests, but we were civil. We couldn't prove that the other guests had been VC, so we couldn't just arrest everyone. It hadn't been that long since the big Buddhist riots in Saigon and Hue. Our superiors would not have looked kindly on any rough treatment of monks, not without very substantial evidence.

All the VC slipped out the back while 11th Company was relaxing in the front. I didn't blame the troops, but I thought that Vinh should have been a little more observant. I also blamed myself, but not very much. What the hell did I know about Buddhist monasteries and rituals? Only partially suppressing my anger, I asked Vinh why he hadn't suspected anything, particularly since some of the younger VC had seemed not to know what they were doing during the prayer session. With a questioning look, he said, "I am a Catholic—I don't know much about Buddhist ceremonies."

6 HEROES

After the fruitless search of the Go Chua pagoda—and learning more about oriental inscrutability—we turned and moved to the southeast. Tenth Company was finding fresh VC signs near the Saigon River. The battalion CP and 9th Company were displacing to the village of Ap Dong. The evidence pointed to a larger-than-normal Viet Cong force there, so the regimental commander was ordering the 1st Battalion to leave Tan Thanh Dong and set up blocking positions north of our location. He, Major Humphry, and Captain Robinson came in by chopper and set up the regimental command post with our people.

Instead of moving to 10th Company, Lieutenant Vinh was ordered to get 11th Company to Ap Dong as quickly as he could. There was something up, and everything indicated that we'd soon be in a big fight. To save time, we cut south to an east-west dirt road and followed it to the far end of Ap Dong, where the battalion had set up its CP. Despite our heavy rucksacks and the mushy ground, it only took us a couple of hours to get to our destination.

When we arrived, I left the company and found the joint command post which had been set up in a peasant's home. When I walked in, the regimental commander, Man, and the three advisors were studying a map on a table. The Vietnamese colonel had learned his profession well from the French and Americans. He was drawing large circles and arrows all over the map with colored grease pencils. If I hadn't known better, I would have thought that he was maneuvering whole divisions.

When there was a break, I pulled Kinsey aside to find out what was happening. He said that from what 10th Company was finding, it appeared there may be as much as a full VC battalion a couple of kilometers north of Ap Dong. The colonel was trying to figure out how to surround the VC and keep them from breaking into smaller units and escaping. Division had a U.S. chopper working the area, trying to spot Charlie. Kinsey told me to stand by and be ready to move with 9th Company as soon as something happened.

Less than fifteen minutes later, Major Humphry's radio came alive. The chopper had been hit by ground fire and had crash-landed. The crew was unhurt and was un-assing the aircraft. They had come down some distance from where the VC had fired but needed help immediately, before Charlie got to them.

Grabbing my rifle but leaving my rucksack, I charged out the door. Likens had already heard the commotion and was on my heels as we ran to the 9th Company. I didn't know where the chopper had gone down, but I knew that Lieutenant Chung was getting instructions over the radio. Likens and I got to the company just as it was moving out. Its men had also left their rucksacks to make better time.

Fortunately, the helicopter was only about a kilometer north, and although the ground was wet, we could move pretty quickly without our heavy packs. We didn't hear any firing, so we figured Charlie hadn't gotten there yet.

The chopper had just come into view when we heard some sporadic fire beyond it—it was going to be a horse race. We couldn't see the crew, but we knew that they wouldn't have moved very far from the downed bird. The closer we got, the more the firing picked up. The Huey appeared to be in pretty good condition, and Charlie was trying to destroy it and kill the crew.

The chopper was to our left-front, and the VC were to our right-front, so we charged straight ahead to put ourselves between the crash and them. The VC shifted their fire and began shooting at us, but 9th Company's troops didn't hesitate. Just as we got into position to protect the bird, a rescue chopper roared in to pick up the stranded crew. The pilot was brave but not too smart. He should have waited until we had taken care of the VC. As he was hovering about ten feet off the ground, Charlie hit his helicopter in some vital spot, and it suddenly dropped hard into the rice paddy. In an instant, the chopper's front doors opened, and the pilot and copilot rolled out on the ground. The two door gunners and a third man leaped out the open side doors and threw themselves in the mud. It didn't appear that any of them were seriously hurt, because they were all scrambling for cover behind a small patch of scrubby trees.

We continued firing at the VC for a few more minutes, and then a couple of gunships showed up. I directed their rockets on the VC, and immediately Charlie's fire slackened, then stopped. This was probably no more than a VC platoon, and it was very unlikely that it could be reinforced because the terrain was too open to aerial observation.

Likens had been keeping Kinsey advised of our situation over the radio, and he gave me the handset. Kinsey asked for an update and I gave it to him: Both crews appeared to be okay, we were between them and Charlie, and we only had a couple of wounded. Kinsey asked if it was possible to bring in another rescue chopper. I told him that if it came in very low from the south, it probably would be okay. When it landed to pick up the two crews, it could use the patch of trees they were hiding in as cover from any Charlie who might stick his head up and shoot. However, it had to time its approach with the gunship rocket runs. We could put out some smoke when the rescue chopper was inbound, giving it some additional concealment.

About ten minutes later a second pair of gunships showed up to relieve the first two. As they began to volley rockets on the VC, a chopper called me to put out our smoke screen. When the dense clouds of purple and yellow smoke began to boil up, a Huey roared in behind us and landed next to the trees. In a couple of seconds the downed crews scrambled on board, and the chopper wheeled around and took off to the south. It was just in time, because the sun was setting—and in the tropics, it wouldn't take long before it got very dark.

Ninth Company was in a dilemma. We'd fired up a lot of our ammunition and were without our rucksacks, which contained our food, ponchos, and other equipment we would have liked to have had. We wanted to sweep the area in front of us for VC (live and dead), but now we would have to wait until morning. Lieutenant Chung radioed for instructions and was ordered to return to the battalion perimeter. Tenth Company, which had just arrived at our location, was directed to secure the two downed choppers until they could be extracted the next day.

When we arrived back in Ap Dong, I left 9th Company, which took up its position on the perimeter, and I went to the command post. Humphry and Kinsey were sitting in chairs listening to a highly animated Captain Robinson. I watched for a bit, and then Mosley whispered that Robinson had been on the first rescue chopper, the one that was still sitting in the paddy. "He's been going on for an hour about how *hairy* it had been and about how it felt to have been in a helicopter crash, etc., etc."

"What crash? The chopper had only been about ten feet off the ground when it settled into a soft rice paddy."

I listened as Robinson described this *big* battle, with *thousands* of VC bullets flying everywhere. I chuckled to myself but didn't say anything. He'd stop every few minutes and complain about his injuries, saying that he felt like he had been "hit with a two-by-four." Finally he stopped talking and wandered over to the other side of the room.

Kinsey spotted me and motioned for me to come over to give him and Humphry a rundown on the action. I told them what had happened and said that Uncle Sam had wasted a perfectly good chopper. Ninth Company had already secured the first downed helicopter and could have brought its crew back with them. Sending in the second rescue chopper had also been unnecessary. As we started to discuss the VC, Robinson interrupted us. He was standing there with his shirt off, asking us to check his injuries. Someone got a kerosene lamp and held it up so we could inspect his back. I didn't see a mark on his body, not even a slight discoloration.

"But it felt like I was hit by a two-by-four."

"Well, if you were, it didn't leave any marks."

That night I discovered where the truly big and vicious mosquitoes lived. I swung in my hammock and fought the bastards all night—but I lost the battle. These suckers could bite you through the material in the hammock and your uniform! Sometime during the night I heard a rifle shot in the distance, but when it wasn't followed by more shooting, I dozed back off.

In the morning I was told that one of 11th Company's soldiers had been shot and killed when he went to relieve a guard on the perimeter. He must have gotten lost in the darkness and accidentally walked forward of the line. One of his buddies naturally assumed that he was a Viet Cong and blasted him. It was too bad, but as we said, "That shit happens."

The rest of the operation was a bust. Sometime during the night, the VC had taken off, and while we located a few tracks, we made no contact. Tenth Company had found one body, several pools of dried blood, and some dirty bandages where we had fought the VC for the choppers. That was the extent of our results for the entire three-day operation.

Back at Paris Tan Quy, we rested up while waiting for the "big boys" to decide on what they wanted us to do next. I noticed that Major Man seemed to be depressed and wasn't his usual smiling self. He had been like this for two days when I asked Kinsey if he knew what was bothering Man. Kinsey told me that Colonel Quan had gotten Major Man disciplined because of the incident involving the soldier killed by the perimeter guard. Quan accused Man of negligence, and the major had gotten a ten-day suspended jail sentence.

I was dumbfounded. I'd never heard of jailing an officer for something like that—a reprimand possibly, but certainly not *jail*. It hadn't even been Man's fault—just an understandable accident. That goddamned Colonel Quan was really a devious bastard!

I held my temper for all of about ten minutes and then went to the TOC (Tactical Operations Center) to find Man. He was slumped in a chair with a long look on his face. This punishment wouldn't end his career, but it certainly showed that Quan was really out to get him. "Hello, Bob. Sit down and tell me a good Vietnamese joke."

"I'm not here to tell jokes. I'm here because I just found out what that fucking Quan has done—and I've got a suggestion for you."

"What's your suggestion?"

"Well, listen to my plan before you stop me. Our recon platoon has AK-47s and black pajamas that it sometimes uses on patrol, right?"

"Yes, go on."

"I suggest that we wait until Colonel Quan drives his jeep to Paris Tan Quy for a visit. Then we use the recon platoon to ambush and *kill* him. Everyone will think that it was the VC, and we'll be rid of the bastard for good."

Major Man stared off into space for a couple of moments and then looked at me and said, "Thank you, but that is a Number Ten idea. We cannot kill our own people."

I argued for my plan for several minutes, but Man would have none of it. He became nervous as I talked and began furtively glancing around the bunker. Realizing that I was making him extremely uncomfortable, I finally gave up. As I turned to leave, he took my hand and held it for several moments. Then he said, "You are a good friend." I walked out, so frustrated that I seriously considered the possibility of ambushing Quan on my own.*

The next morning the battalion formed to move up to Phu Hoa Dong. The headquarters for the Phu Hoa district was located there, and it had been having a lot of trouble with the VC. Phu Hoa Dong was just like Ap Nha Viec—but on a much larger scale. Where Ap Nha Viec held maybe 100 people, Phu Hoa Dong had almost ten thousand. It was right across the river from the Iron Triangle, and many of its residents were relatives of the VC in and around the Triangle. The district headquarters had only one RF company, whose 100 poorly armed and trained soldiers were expected to maintain some sort of government presence. They would make a show of patrolling around the village during daylight and then retreat into their compound at night. The VC had always been able to move freely into and out of the village after dark, but now they were doing it in the

---

*I never completely dismissed the idea of killing Quan during the entire time I was with the 5th ARVN Division. If I thought I would have gotten away with it, I might have.

daytime. The 5th ARVN Division had finally agreed to a request by the Binh Duong province chief to put some regular army soldiers in Phu Hoa Dong, at least for a while. Our 11th Company had been picked for the job, and it required the whole battalion to escort the company through VC country to the village.

Man wanted advisors with 11th Company for the time that it was to be there. However, since Likens was scheduled for R&R, Kinsey decided that it would have to make do with only one advisor—me. This was the first time that I would be operating alone for any length of time, and it bothered me. Two Americans could look out for each other. If one was wounded, the other would see that he was taken care of and evacuated. One advisor was completely dependent on the Vietnamese. If he were incapacitated, there would be no U.S. artillery, helicopter gunships, airstrikes . . . or medevac.

I wasn't concerned about 11th Company's troops, but the idea of losing all contact with U.S. support worried me. Kinsey felt that it would be okay, since there was an American advisory team in the district compound. That was true, but 11th Company was going into the RF compound, almost a kilometer away from district headquarters.

I joined 11th Company and stood around with Lieutenant Vinh, waiting for the order to move out. As we waited, the troops became more and more impatient; everyone was ready and wanted to get going. I was about to ask Vinh about the delay when a chorus of catcalls went up. A soldier was walking in through the gap in the berm, carrying only a rifle. Vinh yelled at him, and the soldier came over with a sheepish look on his face. Vinh uttered a couple of harsh words and then hauled off and knocked him on his ass. He kicked him to his feet and knocked him down again. Then Vinh motioned, and an NCO dragged the battered and bleeding man off.

Vinh explained that the day before the soldier had been given permission to visit his family in Tan Thanh Dong but had been instructed to return no later than 0600. He was an hour late and had held up the entire company.

The Vietnamese Army had a much different view toward discipline than did the American army. It was not uncommon for an officer or an NCO to punch a subordinate and sometimes beat him badly. If a man did something particularly onerous, his head was shaved and he was put in a small barbed wire cage for several days. The typical cage was about two-feet high and about four-feet square—too small to sit up fully or stretch out. The cages seldom had any cover, so the man had to lie in the hot sun or heavy rain for as long as his sentence called for, all the while being fed only plain rice and water. Although some of the NCOs I knew had been a little rough on their troops, this was the first time that I'd seen an officer hit a man.

Finally we began to move up the road toward Phu Hoa Dong, with 11th Company leading and the rest of the battalion right behind. While the road was passable, we weren't using trucks. If we tried to move by convoy, we would be asking for trouble. Even if we weren't ambushed, a truckload of troops made a very tempting target. We had about seven kilometers to go and figured that unless we made contact, it should take us about two-and-a-half hours.

The first four or five kilometers were through rubber trees, probably owned by some Frenchman or a rich Vietnamese now living in Paris or Saigon. Nobody gathered the sap any longer, so there were patches of brush growing wild between the lines of evenly spaced trees. It was eerie, and I kept thinking of all the French, American, and Vietnamese units that had been wiped out in just such a place. I was relieved when we reached the edge of the trees, and I was certain that I wasn't the only one who felt better.

From the trees, the road crossed the rice paddies and then disappeared into Phu Hoa Dong in the distance. The scene was very pastoral, with the sun shining on the green waves of rice and the fluffy clouds floating in a bright blue sky. There was just one problem—there weren't any peasants working the fields. At this time of the morning, there should have been hundreds of them scattered through the paddies.

Major Man deployed the battalion, and we moved toward the village in six irregular files, with each line walking along a dike between the paddies. The battalion had just passed the center of the valley when two or three VC snipers started shooting at us. They fired single shots at a time, but all of the bullets were going over our heads. They were somewhere in the bushes and trees along the edge of the village and were shooting slightly downhill. (Many VC, especially the guerrillas, were poor shots, particularly at long range.)

It was very tempting to call for artillery or gunships, but even if we had known where the snipers were, we would have had to put the rounds on the village itself. Stoically, our soldiers continued to move toward the objective, although at much faster clip than before. I was certain that as soon as one of our men was hit, the battalion would blast the village with everything it had. Possibly the snipers understood that, because after a couple of minutes they stopped shooting and disappeared.

We reformed when we got into the village and marched up the road to the district headquarters, with squads screening our flanks in case the VC attempted something. The people merely stood in their doorways and watched us, showing no emotion. A herd of buffaloes probably would have provoked more interest.

The district compound, which consisted of a couple of small one-story concrete block buildings and several tin sheds, was surrounded by the usual earth berm and barbed wire. The sides of the corrugated tin buildings were sandbagged to about mid-height, providing some protection from small arms and shell fragments but very little from RPG rockets. None of the roofs would stop 60mm or 82mm mortar rounds if the VC decided to drop some into the compound.

I'd been in Phu Hoa Dong several times before, but this was the first time that I had been to the district compound. The American advisors invited me into a tin building and briefed me. Basically, 11th Company was to run squad- and platoon-sized patrols throughout the village and try to put a damper on the VC, who had lately been getting pretty blatant. The RF company would be doing the same thing, plus putting out its usual night ambushes. Eleventh Company's commander would be in charge of both units. If he had any trouble with the RF commander, he was to tell the district chief, who would take care of the problem.

The American major went on to say that he didn't expect much to happen while we were there, because Charlie wasn't likely to start anything that would give us an excuse to use artillery or air on the village. "But don't let your people become complacent, because Charlie will try to pick you off a few at a time. And don't walk down the middle of the main street—you're likely to have some VC sniper shoot your ass. Stay close to the buildings on either side and be careful."

I had a sinking feeling when the battalion left, but it was nothing compared to what I felt when we entered the RF compound. It appeared to have been the headquarters of a rubber plantation, but that had been long ago. The compound was about 200 meters square, with a large house in the center and several smaller buildings around the edge, all needing a lot of repairs. At one time the house had probably been a mansion belonging to a rich Vietnamese or a well-to-do Frenchman. It was two-stories high and constructed of concrete that had originally been painted white. Now it was a light brown from all the dirt and dust. On the ground floor there were patios in the front and back, which had been enclosed with sandbag walls. Each wall had narrow slits, so the defenders could fire without exposing themselves. On the upper story there were covered balconies, with more sandbagged walls. The roof was red terra cotta tile, and perched incongruously on the very top was a small, rickety, sandbagged watch tower.

The RF company commander, a "third" lieutenant (warrant officer), showed us into the mansion. We were immediately hit by an overpowering smell. The troops had been using the dining room as a latrine, and the floor was covered with bits of brown-smeared paper and small piles of shit. Vinh and I backed out quickly, and he angrily told the RF commander that he wanted the practice stopped immediately and the place cleaned and disinfected. We then climbed the staircase to the second floor. Here the bedrooms had large, louvered double doors that opened onto the balconies to take advantage of any breeze that might come by. Whatever furniture or fixtures had originally been in the house had been removed or stolen years before. The rooms were stripped of everything, except the RF's bedding and equipment. The RF lieutenant had his people clean out one bedroom for our company's command post. Vinh and I dropped our packs so that we could continue our inspection without having to carry them. He told his radio operator to stay with the equipment, and he told me in English to always watch my stuff—

"These bad soldiers will steal everything you have if you do not guard it."

Except for the dining room floor and the fact that the house looked like a refugee camp, it wasn't going to be a bad place to stay. It had good, solid concrete walls, which were fortified with sandbags where necessary. I didn't know how well it would stand up to RPG and recoilless rifle fire, but I hoped that I wouldn't find out.

The rest of the compound was a complete disaster. The berm and sandbagged bunkers had apparently been built years before and had obviously never been maintained. Most of the sandbags were rotten and some bunkers were unusable—or the danger of their collapsing offset any protection they might have offered. In several places the RF had cut openings in the berm so the men could take shortcuts to the village. These saved them from walking 50–100 meters to the main entrance, but it completely defeated the purpose of the berm. The several paths through the barbed wire were so well worn that the VC wouldn't waste their sappers on the defenses.*

Many of the RF soldiers' families lived in the compound with the troops, so there were little hootches scattered all around, and a lot of children playing. Hanging laundry, cooking fires, hammocks, and trash made it look more like a squatters camp than a military fort. No one in 11th Company was very impressed or happy, but we'd have to make do for the next ten days.

Lieutenant Vinh called his platoon leaders to a private meeting. He gave them instructions to be prepared to withdraw their platoons to the house for a final defense if we started to get overrun by the VC. From what we had seen, we were not going to rely on the RF in the event of an attack. They could do whatever they wanted—we were going to look out for ourselves.

For the next several days I did very little. Vinh sent two platoons out each day and kept the company headquarters and a platoon in

_____

* Sappers were specially trained Viet Cong who could penetrate the defenses of almost any Allied installation. Once through the barbed wire and minefield, the sappers would destroy command posts, crew-served weapons, and communications facilities with explosive satchel charges. Sappers were the most feared and respected of all Viet Cong soldiers.

reserve in case one of the patrols ran into something. Since Lieutenant Vinh wasn't going out himself, there was little reason for me to go either. Besides, MACV had a rule against a single advisor operating with anything less than a company. I didn't pay any attention to the rule, but I felt that I would be more useful with the company commander should something happen. I was beginning to understand Kinsey's view of his role as advisor to Major Man.

On the second day, I climbed up to the watch tower through the hole in the roof to get a view of the surroundings. I was standing there like a lookout in a crow's nest when a couple of RF soldiers began yelling up at me. I couldn't understand what they were saying, so I just waved at them. A few minutes later an RF sergeant climbed up the ladder and pulled at my leg. I bent down and he breathlessly told me that every time they had put somebody in the tower, the VC shot at him. After a couple of close calls and a couple of wounded men, they had stopped using it. I decided that I had seen enough and climbed down into the attic.

If the VC had shot at the guard that often, it meant that they kept the compound under more or less constant observation. Since 11th Company's arrival was out of the ordinary, we could be pretty certain that they were closely monitoring us now to see what we were doing.

On the third night, I was attempting to sleep but was having a hard time with a bunch of mosquitoes that didn't seem to be the least bit intimidated by my insect repellent. I was just dozing off when I heard someone distantly talking over a loudspeaker. I rolled over, slapped another mosquito, and then heard some whispering. I opened my eyes and saw Vinh and his radio operator kneeling at the sandbagged balcony wall. Crawling over to them, I asked what the hell was going on.

"Lieutenant Parrish, you are a famous man."

"Why do you say that?"

"Because the VC are saying that they will pay one thousand dong [piasters] if one of us kills you."

"Is that the VC out there with the bullhorn?"

"Yes."

"They are only offering *one thousand*? That isn't very much."

"Maybe the price will go up the longer we stay here."

"I hope so. A *water buffalo calf costs a hell of a lot more than a thousand dong!*"

We all laughed, but I was a little upset. Many Americans loved to believe that the VC had put prices on their heads. I'd heard so many people make the claim that I'd stopped putting any stock in their stories. I certainly wasn't going to tell anybody about this—particularly since a thousand piasters was less than $10. The next day several 11th Company soldiers made a point of bumping into me, saluting, and joking that *they* thought I was worth much more than a thousand piasters.

The VC with the bullhorn repeated his pitch the next night, but the price was still the same. On the following night, he had just started talking when there was an explosion. A few 11th Company soldiers had crawled out after dark and had waited for the guy to show up for his nightly show. When he did, they blew him away with an M-79 grenade launcher. Vinh said that he hadn't sent them out—they had done it on their own. That made me feel good, and I dismissed my own plan of paying a soldier to go out, pretend to be a VC, and offer something like fifty thousand piasters. I had been having a hard time coming up with just the right figure. It had to be enough to satisfy my pride, but not enough that someone might try to collect.

I guess we pissed off the VC when we killed their man, because the very next night a squad of them fired at the compound from the far tree line. They stopped when we dropped a few 60mm mortar rounds where we thought they might be. During the firing, Vinh and his radio operator were busy talking to the platoon leaders and the RF company commander, getting information and coordinating our reaction. I crawled over to the sandbagged balcony wall to observe the action and try to get a fix on the VC. As I was squatting there, the radio operator's wife joined me, wearing her husband's web gear and holding his carbine. Shortly after we moved into the compound, she and her small baby had joined us for an extended visit while her husband was in a static location. She was very handy to have around—she cooked our meals, cleaned up after us, and although she was only about twenty years old, she acted like a rather domineering mother. She and the baby gave a homey touch to our otherwise spartan, military existence.

By the light of the mortar illumination rounds, I could see that she had a very grim look on her face. She was fully prepared to defend the command post if need be, and it was obvious that she knew how to handle the rifle. I was a little disconcerted because this was the first time I'd seen a woman actively involved in a combat situation.

The rest of the time in Phu Hoa Dong was just plain boring. I borrowed three pocketbooks from the district team and finished them all. Finally the battalion marched back up the road and took us home. Kinsey said that the battalion had only been running company and platoon operations and that I hadn't missed much. Likens had returned to the battalion and, taking me aside, said, "I checked in with the admin people when I came back from R&R. You want to know what I found out?"

Without waiting for me to ask, he continued, "You're not going to believe this, but that fucking Captain Robinson is getting the Purple Heart and a goddamn *Silver Star!*"

"What did he do?"

"It was for that bullshit with the downed choppers at Ap Dong. The recommendation for the Silver Star said that he had *personally* rescued the crews!"

"Hell, 9th Company had already secured the aircraft and crew before Robinson and the second chopper ever showed up! If that dumb-assed pilot had coordinated with us, he wouldn't have had his helicopter hit. Chung had a squad right with the two crews to provide protection if they needed it. What the hell did Robinson do?"

"Yeah, and I'd like to know when he got wounded. You officers checked him and didn't find anything, did you?"

"Maybe whoever wrote the recommendations saw something we didn't."

"Who the hell would that be?"

"I have no fucking idea."

I had only been back at Paris Tan Quy for a couple of days when Kinsey called the team together to brief us on a new operation. Three battalions from the 5th ARVN Division and three from the U.S. 25th Division were going to surround Phu Hoa Dong and seal it off. Each battalion would hold a portion of the perimeter and be assigned a sector of the village which it was to search; whatever VC weapons, ammunition, and supplies were uncovered would be confiscated. Binh Duong Province headquarters would bring in National Police to check identity papers. The police would arrest anyone without papers, people with suspicious documents, and anyone who was suspected of being a draft dodger or an ARVN deserter.

One ARVN and one U.S. battalion would be inserted by helicopter. One U.S. mechanized infantry battalion and the ARVN 1st Armored Cavalry Squadron would come by ground in their tanks and armored personnel carriers. Our battalion and an American battalion would walk. It was all supposed to be coordinated so that everyone reached the village just before daylight. That meant a night move, something that none of us liked. Everything about the operation was hush-hush, but I was certain that the VC knew the details even before they were passed to our battalion.

That evening we moved out before midnight to allow ourselves plenty of time to get there by dawn. Like the Americans, we weren't particularly skilled in moving at night in a large formation, so we didn't do anything fancy—we just put a battalion column on either side of the road. At the edge of the rubber trees we left the road and made a wide swing to the right through the rice paddies. The north-south road had been designated as the boundary between the 25th U.S. and 5th ARVN Divisions, and we didn't want to stumble into the American battalion coming in from the southwest.

Just before dawn we moved to the southern edge of the village and established a line from the road to the east. Our positions were close together, so no one could infiltrate into or out of the village. While this maneuver didn't cause us much problem, there was a lot of noise and confusion elsewhere, as other units' tanks and armored personnel carriers moved to take up their positions. Adding to the din were helicopter assaults in other sectors of the village. If there was any shooting, I didn't hear it because of all the noise.

By the time the sun peaked over the horizon, every battalion was in position and tied in with its neighbors. I was with 10th Company on our westernmost flank so I could coordinate with the U.S. battalion to ensure there wouldn't be problems linking our mutual defenses. After making certain that Lieutenant Tien and his American counterpart were squared away on who had responsibility for what, I went to check out the rest of our positions.

The troops were digging in just inside the tree line at the edge of the village, giving them a good view of the rice paddies to our front. The sun was still low, and it hadn't begun to get hot, so it was a comfortable walk with only my web gear, beret, and M-16. I wandered down our line, stopping every few meters to bullshit with the troops. They were in a good mood, because they would be in a static position for at least a few days and could get some rest.

I reached our easternmost position, where we tied in with the ARVN 2d Battalion, then turned around and started back. I had gotten to about the center of our battalion's sector when firing erupted all along our line. At least a full company of VC had opened up on us from a tree line about 300 meters across the paddies. The Viet Cong had several machine guns spitting pink and green tracers into our lines, and every few seconds an RPG rocket would roar out of the trees and explode among our foxholes. Maybe Charlie hadn't heard that we had *six* battalions around the village—or maybe he had but knew that we were so strung out that we couldn't mass against him.

I ran along the line trying to find the nearest company command post so I could use its radio to call up the artillery. As I ran, I could see the lines of VC tracers shifting in my direction. They had been firing all along our positions, but now they were throwing a lot of it my way. Two things occurred to me: First, I was the only one up and running, and second, I was the only one wearing a beret. Everyone else was down in a hole with a helmet on. I was drawing attention, and my beret showed that I was probably an officer. I threw myself down behind a tree, yanked off my beret, and stuffed it in my pocket. Soldiers sometimes get caught without their helmets, so being bareheaded wasn't particularly unusual. Wearing a beret was—and it ain't smart to stand out in a firefight.

I crawled back through the undergrowth to the hootches behind me. Once I had a row of houses between me and where the VC's rounds were impacting, I got up and ran to the CP.

Everything was buzzing in the house, with all the radios going as Man got information from his companies and in turn, reported it to regimental headquarters. Kinsey grabbed me and asked what I knew about the action. I told him that at the very least, a reinforced Viet Cong company was firing into the center of our line. The VC had at least a couple of machine guns, several RPGs, and possibly even 57mm recoilless rifles—all of which meant that they were main force and not just some ragtag local-force group. It was a standoff because they couldn't attack across the several hundred meters of open rice paddies, but neither could we. We went over to Man, and I repeated what I knew to him. I called to Likens and told him to get ready to go back with me to the action, but Kinsey said no. For some reason Colonel Quan was at regiment, working up a plan, and Kinsey wanted me to stand by.

In a few minutes Man left his radio and called for us to join him.

Using a map, he explained that Colonel Quan wanted him to have choppers land a couple of platoons beyond the VC's positions and attack Charlie from the rear. Pointing at a spot on the map, he showed us the planned LZ, about 500 meters south of the Viet Cong. A tree line on the northern edge of the LZ would mask the landing from VC fire—if Charlie had not occupied it.

It was a plan for disaster. Our two platoons would be no match for a reinforced Viet Cong company. The VC could merely turn from firing toward the village and wipe out the platoons as they tried to attack across flooded rice paddies. I pointed out the obvious and insisted that the only reasonable course would be to allow the platoons to set up blocking positions in the tree line. We could then use artillery and air to force the VC to withdraw—hopefully into our block. They would be the ones who would have to cross open rice paddies.

Major Man agreed, but he said that he had already argued with Colonel Quan and had been told to "execute the plan, immediately and without change." Then looking at me, he said, "Colonel Quan insists that an advisor accompany the platoons."

"An advisor" meant me. The hair stood up on my neck, not only because it was probably a suicide mission, but also because I sensed that Colonel Quan was trying to get me killed. I looked for Kinsey to say something, but all he really could say was, "Take one of the NCOs with you and—good luck."

Likens and I met the platoons outside the CP, and we all moved to the road, where the choppers were supposed to pick us up. Shortly after we arrived, the helicopters landed, and I ran to talk to the pilot in command of the lift. He pulled his helmet off and leaned out the door so he could hear me.

"Have you been briefed on the situation, and do you know where you are to insert us?"

"Roger, that. Are your people ready to go?"

"Yeah, but there's been a change of plans."

I showed him on the map where the VC were and said that we'd probably have a very hot LZ if we inserted where he'd been told. I told him that I wanted to land *two* tree lines south of the VC rather than only one. I expected that Charlie was covering the first tree line, but probably not the second one. We would have to move an additional 200–300 meters, but at least we'd get on the ground without being shot out of the sky on the approach. I got no objection from the lift commander, and it was a much-relieved pilot who gave me a thumbs-up as I climbed in the back of his bird.

We pulled pitch, and after gaining enough altitude, we swung to the west. The choppers flew for several kilometers and then made a 180-degree turn and dropped so low that we were barely clearing the tops of the trees. We may have taken some fire as we came into the LZ, but if we did, it wasn't enough to bother us. Our real worries would begin when we got on the ground. The helicopters didn't land—they just hovered a couple of feet above the paddies, and we jumped from the skids.

Fifty thoroughly scared ARVN soldiers and two American advisors began wading toward the tree line. When we got to the nearest dike, we climbed out of the paddy and began to run as fast as we could along the dikes toward Charlie. Our first problem was to get to the trees before any VC did.

We made it, but we were completely out of breath. When I had recovered a little, I got on the radio and directed a couple of gunships to start laying down fire on the next tree line. I hoped they could pin down any rear guard the VC may have put there while we charged across the next 200 meters of open area. As soon as the choppers began firing, I gave the signal and we started running along the dikes toward where the rockets were exploding. We intended to get as close as possible before we told the helicopters to cease firing, and we were prepared to take some casualties from their rocket fragments.

We only got halfway across the field when something close to a full squad of VC opened up on us. They were obviously dug in, and the 2.75-inch rockets weren't doing a very good job of making them keep their heads down. We jumped into the paddy and took cover behind a dike running parallel to Charlie's tree line. I had been right—if we had landed in this field, we would have lost a lot of people and choppers before we ever got started. Since the gunships weren't having much effect, I called for artillery. When I radioed the VC's coordinates and our location and distance from the target, they refused to fire until I gave them my initials. We were closer than the "minimum safe distance."

81

Not being in a big hurry to get killed, I decided to let the artillery pound the hell out of the tree line before we made any attempt to move forward. Kinsey called on the radio and asked our situation. I told him that we were pinned down and were trying to eliminate the VC's rear guard before moving to the tree line nearest the main VC positions.

After a pause, he radioed, "My counterpart's higher [meaning Colonel Quan] is insisting that your unit close on Victor Charlie as soon as possible."

"You tell your counterpart's higher to go fuck himself! If he doesn't like how we're doing this, he's welcome to come out here and lead the goddamn attack in person!"

"Roger, understand. Do what you can, but take care of yourselves."

Just about the time I figured we couldn't stall any longer, it began to rain. In moments the sky opened up and hit us with as heavy a monsoon rain as I had ever seen. It was coming down so hard that I could hardly see the tree line only 100 meters away. I ordered the artillery to cease firing, and we began to wade forward, firing from the hip, not really seeing where our rounds were going. Here and there, people went down and didn't get back up, so there were still some live VC in the trees. I yelled and signaled for everyone to get on the dikes, since the sooner we got to the tree line, the better off we'd be.

When we closed on it, we spotted several VC running toward their main position, and we cut down four of them before they disappeared in the rain. Again we stopped to catch our breath. The rain was coming down even harder, and it completely blocked our view of the trees where Charlie had his main force. Because of it, the air force had to call off the airstrike, so I radioed for the artillery to resume firing. We lay there wet and miserable, dreading the final assault we knew we'd have to make. I didn't expect many of us to come through it unscathed. Naively, I still believed that I wouldn't be killed, but I did expect to be wounded. The only question in my mind was how badly I'd be hit.

I don't know how long we waited in that tree line, but finally we had to move forward. We got as close to the impacting artillery as we could, and then I radioed for a check-fire. I would have shifted it somewhere, but there wasn't any place to put it. The only thing beyond the VC's trees was our battalion's positions at the edge of Phu Hoa Dong. When the artillery ceased firing, the only sound we heard was our own shooting and the rain. The VC must be holding their fire until we got close enough for them to blow all of us away at once. At the edge of the last paddy before the trees, we began to yell—not to scare the VC, but to boost our own morale.

They were gone! There were dead VC bodies scattered along the tree line where they'd been killed by artillery and air, but the live ones had pulled out to the east during the heavy rain. They had gone through the trees—toward the Hook—rather than through us. They must have figured it would cost them too much to stay and fight, so they used the rain to conceal their withdrawal. There were no words to describe our sense of relief—it was just too good to be true.

When we got back to the battalion command post, we were tired, hungry, and thoroughly pissed off. The relief we'd felt when we found the VC gone had been replaced by anger. But after a night without sleep and the nervous exhaustion of the operation, the only thing Likens and I wanted was some rest. Around midnight Mosley woke me for my turn on radio watch. I'd slept for about nine hours and felt much better. I spent the next three hours monitoring the radio, listening to various units reporting on small groups of VC who were trying to sneak out of the village. There was only one serious contact—when about a squad of VC tried to blast its way through the perimeter with hand grenades. A couple of American GIs had been wounded, and a couple of VC had been killed.

The next morning we began the search phase of the Phu Hoa Dong operation. Each rifle company sent patrols through its portion of the village, checking identity cards and searching hootches. Besides the patrols, we decided to send part of a company to sweep the area forward of our lines; we wanted to see what we could find and we wanted to try to keep the VC from getting too close to Phu Hoa Dong. We didn't expect any big contact, because by now Charlie knew that we had six battalions around the village. Our five or six supporting batteries of artillery and the availability of airstrikes and gunships should also have impressed him. Mosley and I decided to go with the sweep force.

The sweep was pretty routine, and while we found the prints of Ho Chi Minh sandals on several muddy trails, they were all headed away from the village. We'd been out for about three hours when I spotted something in a clump of coconut trees 100 meters to our left. I signaled for the patrol to stop and get down. I didn't know what it was, but there were some unusual shapes under the trees. I passed the word, and we began cautiously moving toward whatever it was —ready to fire if necessary.

They were VC bunkers. Because the area was so marshy, the bunkers had been built above the ground and camouflaged with coconut palms. It was the palms that had attracted my attention because of the way they were lying. When a palm branch falls from a tree, it normally falls flat. Several vertical branches look unusual, meaning there had to be something under them. Charlie's camouflage probably would fool a helicopter, but ground troops should have been able to tell that something wasn't quite right.

There were three of them, all made of hardened mud and coconut logs. Each was about six feet high and six feet square, with a small entrance on one side and narrow firing ports on each of the other three sides. They were surrounded by low grass, with a well-used trail coming into the site from the south and leading out of it to the north. This wasn't a fighting complex; these were supply bunkers.

The patrol leader put troops out about fifty meters around the place to provide local security, and we began to search the complex. While the others were checking two of the bunkers, I decided to investigate the third one. I walked over and shined my flashlight into one of the firing ports. I could see the shape of something in the corner, but I couldn't make out what it was. It looked like a pack and possibly a rifle. I went around to the entrance and getting down on my hands and knees, I shined my light inside. As far as I could tell, the bunker was empty, except for the pack and what I hoped was a weapon in the corner. I started to crawl through the entrance and then had a strange feeling. I pulled back a little and sat on my heels, trying to figure out what was wrong. I shined the light all around but still didn't see anything but the pack, so I got ready to crawl in—then it came to me.

There was a small wooden board on the ground, just inside the entrance. A board this far away from a village was unusual, and there didn't seem to be any reason for its placement. The VC seldom threw anything away; they made use of almost everything. There were many uses for an eight-inch by three-foot board, so it was very unlikely that they would have just left it laying on the ground to rot. I pulled off my helmet and web gear and laid them with my M-16 next to the bunker. Then taking my K-Bar knife, I carefully poked around the edges of the board. Finding nothing, I laid my head on the ground and very slowly lifted the edge of the plank with the tip of my knife. Under the center of the board I saw a round metal pressure plate, which I recognized immediately as the bottom of a cluster bomblet.

The U.S. military developed a lot of new weapons during Vietnam—some of them effective and some of them ridiculous. One of the new things the air force had come up with was the cluster bomb unit (CBU). A clam shell container popped open after being dropped from a plane, and it released hundreds of small bomblets. When they hit the ground, they exploded, sending little steel balls in all directions. One bomb canister filled a several-hundred-square-meter area with deadly shrapnel. The first bomblet I ran across in Vietnam was shaped like a pineapple, with little metal fins that looked like pineapple leaves. They popped open and stabilized the bomblet's fall, so it hit the ground with its detonating plate down. Another type looked like a baseball with a small circular plastic propeller assembly. This type would begin to spin as it fell. After a certain number of revolutions it was armed, so it exploded the moment it stopped spinning.

Cluster bombs were effective, but a small percentage of them always failed to explode. The one that the VC had buried under the board was one that hadn't hit the ground just right. Charlie figured that when somebody like me stepped on the board, it would go off, blowing me and anyone near me to hell. He was probably right.

In one of my not-so-smart moves, I carefully dug the bomblet out of the ground with the knife. Holding it by the fins, I walked over to the patrol leader and asked him if he would like a piece of pineapple. Neither he nor any of the others thought that it was very funny, so I went back to the bunker to collect my souvenir. It was a completely rotten VC pack filled with dead grass. The "weapon" was a small tree limb, very roughly carved to look like a bolt-action rifle in the dim light. It had all been a setup to lure someone into blowing himself up.

When we got back to the battalion CP, Mosley told Likens what had happened. Only half joking, he said, "Sir, you ain't going to make it through your first six months at the rate you're getting into the shit."

We spent the next several days patrolling the village, searching and checking ID cards. None of our patrols made any contact, and only a few VC suspects were collected. As far as we were concerned, our part of the operation was next to useless. Charlie had stopped trying to get into or out of the village in our area because the terrain was too open, so we were little more than bored perimeter guards.

As part of the master plan, it had been decided to permanently station one 5th ARVN battalion in Phu Hoa Dong. It was a long overdue recognition of the fact that one 100-man RF company wasn't very effective in a ten thousand person village. The question we asked was, "What battalion would get the new compounds?" The betting was about evenly divided between our battalion and some other unit. The 3/7 had been out humping and fighting the VC for many months, while the troops in the regiment's other three battalions had been mostly sitting on their asses, responsible for only relatively small areas. Because of the nearly continuous operations, our casualties had reduced the battalion's fighting strength to less than fifty percent. The battalion was due for its own compound and small AO, even if it *was* in the worst village in the province. The argument against our chance of being picked was simply that Colonel Quan didn't like us.

We didn't have to wait very long for the answer. One of our sister battalions would take over Phu Hoa Dong, and we would return to the field when this operation finished. Everyone's morale dropped, and our companies began reporting a few desertions.

A couple of battalions pulled out, and we were shifted to the north end of the village, to just across the Saigon River from the Iron Triangle. There the VC were still crossing the river at night and going into and out of the village. The main road coming up from the south from Paris Tan Quy passed through the center of the village and dead-ended at the river. We were given responsibility for the northeastern quadrant, and a mechanized infantry battalion from the 25th Division had the northwestern quadrant. As before, the road was the boundary between the two battalions.

While it may have calmed down in the southern part of the village, things were still hopping up here. Every night at least one of the ambushes would make contact, and my people were reporting two or three kills each time. Since it was officially forbidden for an advisor to go out with anything smaller than a company, we didn't see the bodies and couldn't confirm the kills.

Early one evening 11th Company opened fire on what it said was a VC squad. Shortly before dark a squad of soldiers came to the battalion CP and the squad leader reported that his men had killed six Viet Cong. I stood on the porch of a small house and listened to the excited men describing the fight to Major Man. Man thanked them, saluted, and walked back into the house. As the soldiers were turning to leave, I stopped them and started questioning them on the details. I was tired and not in the best of moods, so I told them that I wondered if the patrol had really killed six VC—we couldn't report any confirmed kills unless we had proof.

"What kind of proof do you want, Lieutenant?"

Without thinking, I gestured toward the neighboring American positions and said, "If they can't bring the bodies in, at least they can bring the ears back."

The next morning a soldier came to the door and asked to speak to me. I went out on the porch to see what he wanted. Saluting smartly and grinning, he handed me six bloody ears. I didn't know what to say, because although some American GIs did take ears, it was strictly against the law. If it got back to the army that an American officer was instructing his Vietnamese to cut ears off VC bodies, I'd be professional "dead meat."

Handing the ears back to the man, I said that I had been joking. I believed our people when they reported kills, and there was no need for them to prove it to me. However, if they still wanted to show me proof, something like a VC pistol belt or ammunition pouch would be sufficient. I went on to say that if we mutilated

VC bodies, they would do the same to ours. They might even start cutting heads off. He thought about it for a moment and then said, "Okay, Lieutenant."*

There are always coordination problems when two units tie in their defenses, as we and the American battalion were doing. While the road was an ideal dividing boundary, each battalion was concerned about what its neighbor would do. To avoid problems, we had both pulled back from the road a little, leaving a gap of about 100 meters between us. Since the road ended just a short distance away at the river, this part wasn't used very much, and the brush had grown until it was little more than a foot path. It is military doctrine that the weakest point in an enemy's defense is the boundary between two units. Besides being doctrine, it is common sense, and both the Americans and we were concerned that the VC were infiltrating our positions along the road. Two days after we moved into the area, I went over to work out the problem with the American company commander. We both agreed that we could put a crimp in the VC's style if we set up an ambush along the road at night. We also agreed that my people would be more effective, because they were "leg" infantry and because they were Vietnamese. So that night, Major Man sent out a squad to try to kill some bad guys.

Sure enough, shortly after midnight five or six VC came bopping down the trail and straight into our ambush. In a minute or two there were three dead and one captured VC. The captured Viet Cong had been shot in the leg and was brought blindfolded on a stretcher to the battalion CP. I tried to interrogate him, but he wasn't very cooperative. It really didn't matter though, because all he would have told me was that his unit was in the Iron Triangle. After our medics

---

*Cutting a head off was pretty much the ultimate a Vietnamese could do to his enemy. If the head was not buried with the body, the dead man's soul could not rest, and it would wander through eternity. There were complicated rituals that could be performed to "capture" the spirit in a jar, so the deceased could come back and live with his ancestors, but they were expensive, and there was a risk that they might not be successful. I never saw the Viet Cong or the ARVN cut any heads off. No one wanted it done to them, so they didn't do it to the other side.

88

gave him first aid, they loaded him in a jeep and took him to district headquarters.

Later in the afternoon, I found one of the VC bodies *standing* against a barbed wire fence, with a cigarette in his outstretched hand. As was our standard practice, we had left the bodies along the road so the villagers could see them and collect them for burial. After rigor mortis set in, some Americans had propped up one of the bodies on the fence. He hadn't been shot up too badly, so when I first saw him, I thought that he was alive. I knew that American GIs had been responsible because of the cigarette. It was a U.S. brand, and none of my people would have wasted it. They might have used one of their own Vietnamese cigarettes, but certainly not an American one.

I was getting bored hanging around the battalion CP, so I decided to spend some time with one of the companies. I chose 10th Company because Lieutenant Tien was our best company commander and I liked working with him.

Tien took the whole company on patrol near the river, figuring that if we were going to make contact, that would be the most likely place. We ran across several formerly well-to-do houses that apparently had been owned at one time by middle-level rubber plantation managers. Most had now been taken over by peasant squatters who were too poor to maintain them. Early in the afternoon Tien called a halt next to one of these houses and told the company to set up for the night. The company command post and two platoons would stay here, while the third platoon would send out three squad ambushes.

The soldiers who doubled as the headquarters' cooks built small fires and began to fix dinner. The house was locked up so I looked around to find a place to string my hammock and poncho. I tied one end to a low decorative concrete wall along the edge of the veranda, and the other to a porcelain knob on the front door. I tried the hammock and it seemed to be sturdy enough. I didn't have any place to hang the poncho so I decided just to drape it over the hammock to keep the rain off me.

Lieutenant Tien and I sat on the veranda and bullshitted well into the night. Finally, about 2200, I decided to get some sleep and crawled into my hammock. I'd just gotten comfortable when the damned

89

doorknob broke and dropped me on the hard concrete. Swearing silently to myself, I crawled around in the darkness, trying to find someplace to rehang it. There wasn't anything to tie it to, so I finally said to hell with it and curled up on the concrete next to the porch wall.

I was sleeping soundly when there was a burst of fire and I was showered with small chunks of concrete. I grabbed my rifle and slid down the porch wall to the opening, ready to return fire. There was some yelling, a few bursts of fire, and a couple of hand grenades thrown by our people, but nothing more from the VC. I only dozed the rest of the night.

When I woke in the morning, I looked up on the house wall. There were a dozen bullet holes in the door and on the wall, exactly where I had strung my hammock. I called to Tien and showed him the marks. Then we went forward of our lines to search. About fifty meters beyond our outer positions, I found two forked sticks stuck in the ground in some bushes. I laid my rifle in the forks and sighted down the barrel—*exactly at the spot where I had hung my hammock*. Some VC had watched me string it and had put aiming sticks in the ground. After midnight he came back, put his AK-47 in the forks and fired a burst. Had the door knob not broken, my perforated body would have been already wrapped in a shroud, ready for evacuation to Graves Registration. I learned another valuable lesson—*never* set up your position for the night before it gets dark. Pick it, but don't go to it while the bad guys might be able to see you. So far, I had been incredibly lucky learning these lessons before it was too late.

The next day I rotated to 11th Company, which was doing essentially the same thing as 10th Company. On this patrol, we discovered a hootch with fifty or sixty Ho Chi Minh sandals hidden under some rice. Although we had the woman who lived there cold, she refused to tell us anything. Lieutenant Vinh was in a quandary about what to do. Normally, he wouldn't have hesitated to bang her around to make her talk. (We advisors argued against mistreatment, but many Vietnamese ignored us.) The only reason he didn't do it this time was because the woman was very pregnant. Vinh pondered how to handle the situation and then appeared to have a bright idea. I made my obligatory protest and moved off to the platoon that was the furthest away. Although I couldn't stop whatever it was he intended to do, I was not going to stay and be a passive participant.*

As I leaned against a tree, I heard loud feminine laughter coming from the hootch. There would be a pause, and then the laughter would start again, louder than before. I was totally confused, and the platoon leader with me had no idea what was happening. These bouts of laughter went on for about thirty minutes, and then there was silence. Vinh emerged from the hootch and I walked down the trail to join him. He told me that they had gotten what they wanted from the woman by tickling her with a goose feather. For some people—obviously including her—tickling is a form of torture. I thought that this was a very original idea that the Geneva Convention could never argue with—we'd tickle the enemy to death.

The woman was Viet Cong, and her husband was across the river in the Triangle. He and some of his buddies would swim the river with inner tubes; he would visit his wife and pick up the sandals that she and her friends made. As it happened, he was due that night.

The hootch was isolated, so Vinh was pretty certain that no one had seen us arrest the woman. He set up his command post in the house (with the woman tied and gagged) and put a ring of troops around it about fifty meters out and well camouflaged. About 0100 the husband and two other guys walked up to our positions, and we blew them away. The woman never shed a tear, and the next day we burned the hootch and turned her over to district headquarters.

When I got back to the battalion command post, Major Man met me, and with a sad look on his face, he handed me a large manila envelope. I opened it and pulled out a certificate—the Gallantry Cross for the "Du Ma Ho Chi Minh" thing in the Hook. Then I looked at it more closely. In big printed letters it said, "GIAY BAN KHEN," which translates as "CERTIFICATE OF COMMENDATION." I could speak Vietnamese better than I could read it, but with a little help, I translated the citation:

* Advisors had no authority over the Vietnamese. There was little we could do if they chose to disregard our advice.

The Commanding General, 5th Infantry Division, highly commends Robert D. Parrish, 1LT, OF107329, U.S. Army, who is an exemplary officer, and who has always displayed fierce anti-communist spirit. In a combined operation at XT745195 (Binh Duong) on 4 July 1967, he contributed his brilliant ideas to the 9th Company Commander on the operational plans as well as his fervency in making efforts to get transportation means in evacuating the wounded to the rear for timely treatment, thus creating deep impression towards [sic] the Vietnamese troops.

Brigadier General Pham Quoc Thuan
Commanding General

This was *not* a Gallantry Cross; it was only a Certificate of Commendation, the kind that was normally given to low-ranking Vietnamese soldiers. To give one of these to an officer, particularly an American officer, was an insult. The Vietnamese knew this well and knew that for an officer, getting no award was better than such a certificate.

With genuine sadness, Major Man apologized. He had recommended me for a real medal, but someone above him had reduced the award to this certificate. He fumbled with his words so much that I stopped him, telling him that it was no big deal. He wasn't to blame, and I still appreciated him for recommending me for a medal. When Man left the room, Kinsey took me aside and said that Colonel Quan had made division downgrade the award. Although I already knew the answer, I asked him why.

"Quan has spies in this battalion, and one of them reported that you had recommended killing him. Be very careful and watch your back."

8 CHANGES

The Phu Hoa Dong operation finally ended, and we marched back
to Paris Tan Quy to return to our search and destroy routine. We
had definitely disrupted the VC, but I didn't think that we had done
any real long-term damage. The village would continue to be a thorn
In the government's side for a very long time.

When we got back, Kinsey called us together for an announcement.
He only had a few months left in country, and it had been decided
that he and Captain Robinson would switch jobs. Kinsey would become the regimental staff advisor, and Robinson would take over as
our battalion senior advisor. This didn't go over well, and there was
a lot of cussing and swearing. A four-man team such as ours was like
a small family. At times we got on each other's nerves—but in the
same way that brothers did. We sometimes complained about Kinsey,
but we also complained about each other. We knew that if we weren't
killed or wounded first, we'd all leave eventually, but Kinsey's un-
expected transfer came as a shock.

Losing Kinsey was bad, but as far as Likens and Mosley were
concerned, getting Robinson was worse. They didn't like the fact that

93

he'd gotten the Silver Star and Purple Heart for the downed heli-copter, and they didn't want to work for him. I wasn't certain how I felt, but at least on the surface Robinson seemed okay. I hadn't had much contact with him, but he always listened to whatever I had to say and never pulled the "captain-lieutenant" routine on me. He recognized that my four-plus months in the bush gave me a lot of credibility and entitled me to some respect, despite my rank. There was some talk about requesting transfers to other units, but I finally convinced the others to give him a chance. A rumor that the battalion would soon be going to Nha Trang for refresher training helped my argument.

Once the decision was made to change the two officers, Division didn't waste any time—Robinson arrived two days later, and with good news. In a week we would return to Phu Van, pick up a large number of new replacements, and fly to a training center north of Nha Trang. Meanwhile, we had one minor operation to run near Tan Hoa, along the western edge of Binh Duong Province. We'd spend a few days sweeping the area and then move back to Phu Van.

Tan Hoa didn't have much of a reputation for anything. Its sole claim to fame was that it was almost exactly halfway between Paris Tan Quy and Cu Chi. From what I'd heard, it was more like Tan Thanh Dong than Phu Hoa Dong—meaning that it was pretty harm-less. It seemed a waste of effort and was probably only meant to keep us busy until it was time to go to the training center.

We left Paris Tan Quy early in the morning, moving south through the paddies. After about three kilometers we hit the stream that served as the provincial boundary between Binh Duong and Hau Nghia. Then we turned northwest and followed the stream toward Tan Hoa. Normally I would have moved with one of the rifle companies, but since Captain Robinson was new, I stayed with him to explain how the battalion operated and what the advisors did. Just after dark, we decided to set up in one of the little villages near Tan Hoa. It was later than we normally went into our night position, but this was a quiet area, and we hadn't seen any sign of the VC.

About 2100, Major Man, Robinson, and I were sitting in a peasant's hootch, discussing how the battalion worked, when suddenly we heard firing along the perimeter. It sounded as though we'd just been hit by a VC company. Major Man immediately grabbed his radio to find out what was happening. I looked around for ours, but it had disappeared—and so had Captain Robinson. At first I thought that he'd gone outside, but then I heard him call for me to get in the hootch's bunker. Looking under the bed, I found him in the hole, trying to contact Division on the radio; he yelled at me that I'd better get in there with him before I got my ass shot. I started to argue that we couldn't call artillery on the VC if we were hunkered down in a bunker, when Likens thrust our second radio in my face. I didn't have time to be debating with Robinson about where advisors were supposed to be in a fight, so I grabbed the one in Likens' hand and charged outside.

We had moved into the village after dark, and apparently the VC hadn't known we were there. When the Viet Cong company walked up to our perimeter, its men must have thought they had run into a small ambush. When they tried to flank the "ambush," they became targets for more of our people. Instead of realizing their error and pulling back, they kept moving parallel to our line, looking for the end—not comprehending that we outnumbered them and had a perimeter completely around the village.

I called an American artillery battery at Cu Chi and adjusted its fire behind the VC. Now they were really in a trick. They either had to try to come through us, or withdraw into the exploding artillery. Since we were putting out so much fire, they chose the artillery.

After the VC broke contact, I went back into the hootch. Robinson slowly climbed out of the bunker, self-consciously laughing and talking about the fact that this was his first time in combat. I gave him a cold look.

It was a good thing that it was late, because we could all turn in. Everyone was so embarrassed that no one wanted to say anything. The next morning we searched outside our perimeter and found a 300-meter swath of VC bodies, weapons, and blood trails. We only had six people wounded, and everyone agreed that it was a good night's work.

I made certain that I stayed with the lead company on the march to Phu Van, because I wanted to be as far away from Robinson as possible. Both the NCOs wanted to go with me, but one had to remain with the battalion command group.

During the first two days back at Phu Van, I made myself as scarce as I could. Finally, on the third day, when the NCOs were gone, Robinson asked me to sit down and talk. "Bob, you were great the other night, and I think you deserve a Bronze Star for getting in the middle of that battle and adjusting artillery."

"What was so brave about that? That's what we're paid to do."

If Robinson had known me better he would have ended the conversation right then, but he didn't. He continued explaining why he thought I deserved a medal. Finally I blew up. "The day after tomorrow we go to the training center for six weeks. Some of the Vietnamese officers know the place and tell me that it is a totally pacified area with no VC. You can play senior advisor while we're there, and then I expect you to get yourself reassigned out of the battalion when we come back here. If you don't, I will personally go to the Division SA [Senior Advisor] and tell him that he has to transfer either you or me."

He said that was fine with him. I figured I could gut it out for another month.

Kinsey and Robinson weren't the only personnel changes we had . . . Colonel Quan got Major Man relieved of command. If there was an official reason, no one seemed to know what it was. One moment Man was the battalion commander, and the next he was gone. The battalion executive officer, Captain Phat, would take over until a new commander reported in. Everyone in the battalion was upset that we had lost the strong and competent Major Man. I hoped that some of our people would take revenge and blow Colonel Quan away, but apparently they were afraid of him, because nothing happened. I guess I didn't blame them, because I was a little afraid that Quan would figure out a way to get me.

During this time of turmoil in the battalion command group and the advisor team, the rifle companies were busy absorbing more·than 200 new troops. They had just completed basic training, and none of them had ever been in combat. With this many new recruits at one time, we were happy as hell that we'd have a month at a training center to get them in shape. The battalion would be in a real hurt if we just charged off into the bush with almost half our people inexperienced.

The two-day move to Nha Trang was an administrative and logistical nightmare, but when the last aircraft finally landed, everybody was accounted for. We had hoped the training center would be near the South China Sea, but we weren't that lucky. It was back in the Annamite Mountains, over forty kilometers from Nha Trang's beaches, bars, and bikinis. The area was pacified—and very boring.

When our convoy pulled into the training center, the American driver suggested that we go meet the senior advisor and then move our stuff into the advisors' compound. The word "compound" was completely inappropriate for where the training center advisors lived. "Country club" would have been a better term. From the fence next to the main road, a manicured grass lawn with small palm trees and flowering plants stretched up to a large whitewashed villa. Most of the officers and NCOs had rooms in adjacent buildings—next to the tennis and volleyball courts! It was almost as good as Saigon.

The driver pulled the jeep up in front of the villa's main entrance, and I half expected a butler to meet us at the door. The inside was just as impressive as the outside. Pictures on the wall, nice furniture, waxed floors—the whole works. The full-colonel senior advisor welcomed us, and after a short, one-sided talk, he told the driver to show us to our quarters. The whole damned place was just like being back in "the world."

The next day the deputy senior advisor (a lieutenant colonel) briefed us on the training program and the rules and regulations. The training his people planned was pretty much standard stuff, but some of the rules were a bit odd. All Americans were *required* to carry their weapons anytime they left the advisor compound. (This was Vietnam?!) No one was authorized to leave the valley where the training center was located, except to go to Nha Trang—and under no circumstances was anyone to go past the 47-kilometer marker on Highway 21. It marked the limit of the training area, and beyond that point was Marlboro Country. A platoon from the nearby Ranger training center had responsibility for guarding the advisor compound, but of course American personnel would assist in its defense if the occasion ever arose. (I asked, but no one could remember such an occasion.)

A few days after we had arrived we got our new battalion commander, Captain Ninh. Ninh, who was about forty years old, was a short, personable guy, given to broad smiles and a bit of bullshit. When we were introduced, he kept shaking my hand and saying that he had heard many, many good things about me. He was a little less effusive with Robinson. Nha Trang was his home town, and his wife, daughters, and mother lived there. He insisted that he would introduce them at the first opportunity.

We didn't plan to spend much time with him for the first week or so, because we felt that he ought to have time to get squared away with his new battalion. However, he had other ideas. The second day he asked me to join him for lunch at a restaurant in a nearby village. It was a little unusual, because he hadn't invited Robinson and he had asked me to keep our little lunch secret. Over several bowls of Chinese noodle soup (out of which one had to pick the drowned flies) and a couple of bottles of "33" beer, he asked my opinions on everything about the battalion and everyone in it. I told him honestly what I thought, including the fact that that while he had two outstanding and one adequate company commander, his staff was a little weak.

After we had talked for an hour about the battalion, he lowered his voice and asked me what I thought of Robinson. I laughed and told him not to worry about my captain. He was only going to be with us until we finished the training and went back home. He broke out in a broad smile when he heard that and said, "Number one! I want *you* to be my *real* advisor, but we will pretend that Captain Robinson is, so you will not get into trouble."

The first couple of weeks was interesting and relaxing. After Phu Hoa Dong, the Hook, and the rest of those places, it was great not to be constantly on the move, always anticipating a firefight or ambush. However, after a while I was going out of my mind with boredom. My sole job was to drive down the road to the center and stand around watching my troops being given refresher training.

Other than the sounds of the training, the valley around the center was peaceful, and it was easy to forget that this was Vietnam. The ARVN center commander had probably made an agreement not to bother the VC and not to let his troops leave the valley. It was also possible that he was paying the Viet Cong to keep the peace. The proof of the arrangement was that Charlie had a tax collection point on Highway 21, somewhere beyond the 47-kilometer marker. Everyday the VC would stop the civilian vehicles and charge a toll to drive up to Ban Me Thuot or down to the coastal highway. Everyone knew of the tax collection point and knew that it was usually placed somewhere around the 50-kilometer marker. I had been told about it by one of the American advisors who was trying to impress me with how dangerous the mountains surrounding the valley were.

One day I mentioned the tax collection point to Lieutenant Tien, and he became angry. Our battalion and others like it were out humping the bush looking for VC to fight, while these training center people probably paid the Viet Cong not to bother them. When Tien had worked himself up enough I suggested that we do something about it. Instead of screwing around watching the training, he and I should sneak up the road and blow the tax collectors away. He broke out in a big grin, and we immediately started making plans.

Early the next morning I picked up Tien at the battalion headquarters. Both of us carried our full combat gear, except helmets and rucksacks, and Tien had scrounged an M-79 grenade launcher to give us a little additional fire power. We drove the jeep to the 47-kilometer marker and parked it next to a bridge over the small stream that defined the training center's boundary. Tien had checked with a couple of civilian drivers the evening before, and they had told him that the tax collection point was between the 50th and 51st markers. The drivers said that there were about six to ten VC and that they moved the point every day or two, but it was almost always between the 50th and 55th markers. We would have liked to have waited until a vehicle came down the road so we could get an exact fix, but none could be expected before noon because of the long drive from Ban Me Thuot. We decided that we couldn't wait that long, so we parked the jeep in the bushes, well away from the road, and moved into the jungle.

We climbed until we were about 200 meters above the road and then turned west to parallel it. We couldn't just walk right up the highway, because we would have been seen by the civilian travelers, and at least one of them would notify the VC that we were coming. Likewise, we couldn't move too close to the road, or we would just stumble into them. It was difficult to see the road from the route we were taking, but we figured that we'd hear the VC stopping the vehicles to collect the money. We kept count of our paces, and after we calculated that we had come about two-and-a-half kilometers, we eased down closer to the road to see if we could see anything. There was nothing, so we moved back up the hill and continued on. I was surprised at how little traffic there was. About every half-hour some truck or bus would go by, but there wasn't a constant stream of vehicles. After another 500 meters, we sneaked down to check out the road—but still nothing. I had brought a pair of binoculars, and I could just make out the "50" on a marker—we should be close.

We continued our pattern of moving 500 meters, creeping down to check out the road, and then moving up and on for another 500 meters. This took a long time, and it was midafternoon before we spotted the 55-kilometer marker. We'd covered the five kilometer stretch that the VC were supposed to be in, but we hadn't found anything. Frustrated, we decided to move along the edge of the road for another kilometer. If we were spotted by a civilian vehicle before we had gone that far, we would turn around and come back. We crept up the road and carefully looked around the bend—but still nothing. We sneaked up to the next bend—and again nothing. Then a fucking bus came by, and twenty or thirty people stared out the windows at us. That was it; we'd really be dumb shits if we continued on.

Tien and I were both unhappy. We'd come all this way and would have to go back empty-handed. No one would ever know we had been there, and the Viet Cong would continue collecting taxes, oblivious to our efforts to kill them. Like a couple of petulant boys, we decided that we at least had to let the VC know we'd been there. We each fired a couple of magazines up the road and pooped a half-dozen M-79 grenade rounds in the direction we thought the VC were.

Then, laughing and giggling, we turned tail and trotted down the road. After a couple of hundred meters we stopped to catch our breath next to a patch of dried elephant grass. We set the grass on fire, so maybe VC who hadn't heard our firing would see the smoke. It caught quickly, and we had to run like hell to get away from it before we were trapped. The walk back down the mountain road was easy and quick, although we were careful to check around each bend in case we had inadvertently bypassed the tax collectors.

I dropped Tien off at the battalion and drove on to the advisor compound. After taking a shower and changing clothes, I went into the officers' bar and ordered a beer. Shortly, several other officers came in to get drinks before dinner and to bullshit a little. One of them said something, and they all went over to the large bay windows that looked out over the valley. I looked up in the mirror over the bar and saw the reflection of the mountains to the west. I could clearly see a large plume of brown smoke rising from the canyon in the far distance. The little fire we had lit had really grown, and it looked like the whole hillside was burning. I should have kept quiet, but like a fool I told the officers what Tien and I had done. The next thing I knew, I was standing at attention in the training center colonel's office.

"So, Lieutenant, you and one of your Vietnamese officers decided to disobey regulations and go off to run your own little war."

I really hadn't expected to get my ass chewed and was a little stunned by the colonel's anger. It looked like I was in big trouble, but I had a few things going for me, and I decided that the best defense was a vigorous offense.

"No, sir. We weren't running *our own* war—we were running *the* war. You, I, and the rest of us were sent to Vietnam to kill Viet Cong. I was doing what I thought I was supposed to be doing."

"No, goddamn it! You were sent here to train your battalion."

"No, sir—*you and your people* were sent here to train my battalion. I was sent here with no change in my orders. I'm still in the business of killing VC."

Fuming, the colonel continued, "I should report your actions and your attitude to your division senior advisor."

"Thank you, sir. I would appreciate that. My colonel thinks the same way I do—so do all my superiors, including General Westmoreland."

Then, going for the throat, I said, "I'm not certain that many people are going to understand how it is that your Vietnamese counterpart *pays* the Viet Cong to leave your people alone."

I had no proof that the training center commander was really paying the VC, but given the graft and corruption that senior Vietnamese officers were involved in, I was confident that everyone would believe me if I made the accusation.

Instantly, the colonel's whole attitude changed, and I knew that I had said exactly the right thing to save my ass. He told me to stand at ease and then proceeded to tell me how he *really* liked aggressive young officers. Sometimes they were a little shortsighted, but that could be forgiven because they took the initiative. He was now acting very friendly and fatherly, but his eyes told me that he was still thoroughly pissed. Knowing that I had pushed it as far as I could, I switched into my modest and respectful mode. We both realized that it was time to terminate the discussion and that the less said about this incident in the future, the better.

Captain Ninh's attitude toward the VC tax collection business was completely the opposite of that of the American colonel. He thought that what Tien and I had done was great, even though we hadn't killed anybody. He pounded me on the back and insisted that he had to introduce me to his family.

The next Sunday, Ninh and I loaded into his jeep and drove into Nha Trang. He took me to his house in a well-to-do neighborhood and ushered me in to meet his family. Ninh's wife spoke better English than he did and was very gracious. She was well made up and wore obviously expensive jewelry. His mother was nothing like the old grandmas that I had seen in the villages. She wore even more jewelry than her daughter-in-law and clearly perceived herself to be the matriarch of the family. She made certain that she was included in all the conversations on more-than-equal terms.

Ninh's wife brought their daughters into the room to meet her husband's advisor. The youngest daughter was about nine years old and very shy. The middle girl was a teenager who repeatedly giggled while covering her mouth with her hand. The oldest one was in her early twenties and was a knockout. Young Vietnamese women are some of the prettiest in the world, and this girl topped any that I had seen. She attended a private Catholic girls' school and spoke fluent French and somewhat better than classroom English.

I didn't know exactly what to say because my previous training had been geared to the more mundane business of war. No one had bothered to tell me how to act in polite Vietnamese society. I decided not to comment on how beautiful the girl was, and I confined myself to a formal Vietnamese introduction. Ninh glanced at the older women and they returned a look of approval. Maybe I was a good guy instead of the western barbarian they had probably expected.

After a pleasant lunch, Ninh suggested that they show me around Nha Trang. His mother stayed with the two younger girls, while his wife and eldest daughter joined us. We took the family car (unusual for a lowly paid Vietnamese captain) that reminded me of a 1950 Ford. Ninh and I sat in the front seat, while Madame Ninh and the girl sat in the back. He drove us up a hill on the western side of the town, where an enormous white statue of Buddha sat. The thing was so big that it could be seen from almost anywhere in Nha Trang. We walked around the statue and looked out over the town, the beach, and the South China sea. If you ignored the grubby tile and rusting tin roofs of the buildings below, it was a beautiful sight.

We got back into the car and drove down the hill into the town. On one of the streets, he stopped in front of a business and explained that it was his wife's tailor shop. When we went inside to tour the place, I noticed there were four or five young men working there who probably should have been in the army.

Madame Ninh was very proud of her sewing machines and took great pains to explain how they worked. Of course *she* didn't do any sewing herself, and she had a manager to run the operation. After leaving that place, we went to another of her businesses, a motorcycle repair shop. If that wasn't enough, we had to drive by a large jewelry shop that Ninh's mother owned. This was no poor army family, barely scraping by on government pay.

It was getting late, so Ninh drove home and we exchanged the car for his jeep. On the way back to the training center, Captain Ninh asked what I thought of his family. I told him that they were very nice and that I was pleased that he had taken me to meet them. Then he asked what I thought about his eldest daughter. Carefully choosing my words, I told him that she was very sophisticated, very educated, and a pretty girl (putting the "pretty" part at the end of the sentence). He broke out in a big grin and said that she and the older women thought that I was a very nice person. I laughed and said that they should see me when I'm killing Viet Cong—which is what I normally did in their country. After chuckling, Ninh said that his wife and the daughter would be visiting us next weekend. I let the remark pass, figuring that he meant that they would be visiting *him*.

Sunday morning the Ninhs showed up at the advisor compound and announced that *we* were going to have a picnic. (A picnic in Vietnam?) The young lady was dressed in a pure silk, pale blue traditional *ao dai* dress, with just enough gold jewelry to show that she wasn't some peasant girl in her best Sunday outfit. Madame Ninh had no such problem. She was outfitted in a very expensive gold embroidered *ao dai* and dripped with gold chains, bracelets, and jeweled rings.

This time I was put in the back seat with the girl and had a very uncomfortable ride, but not because of the bumpy road. I chatted a little with the girl, mostly about how poorly I spoke Vietnamese and how limited my French was. However, I tried to keep a conversation going with Ninh and his wife, making certain to make periodic remarks about my wife and small baby.

We had our picnic at a hot spring, where the water flowed out of a rock. I had brought my miniature camera, and I took pictures of everyone to send home.*

The following day Ninh asked what I *really* thought about his daughter. I told him that she was very nice . . . and that my wife would like her. This last part seemed to pass right by him, and he continued by telling me that all the women in his family thought that I was something special. They liked my looks, my manners, my ability to speak Vietnamese—and on and on. Then he said that his

---

*My wife later wrote, commenting on how beautiful the girl was, but the tone of the letter was "keep your hands to yourself."

daughter particularly liked me and wanted to do more with me. *Oh shit!* Everyone except me seemed to ignore the fact that I was a happily married man, somebody with a family in the States. I began to worry that we weren't going to get away from Nha Trang soon enough. Fortunately the battalion was going to be busy over the next weekend, so Ninh couldn't schedule another family outing.

Thus far, Ninh had shown that he was a competent battalion commander and seemed to be getting along well with his subordinates. Several officers told me that they missed Major Man but thought that Captain Ninh would work out. No one seemed to know much about what he had done before taking command of the unit. As far as they were concerned, he just appeared out of nowhere. It was obvious that Colonel Quan had either picked him or at least personally approved him, so I was very careful about what I said about either Man or Quan. When Quan was mentioned, I made certain to maintain a suitably neutral—inscrutable—face. Ninh didn't try to pump me on Quan, so maybe he didn't know about our poor relations.

One morning as Robinson and I were getting out of the jeep, I saw a prisoner with a shaved head being marched along, guarded by a tough-looking sergeant. I didn't recognize him, but with his two black eyes, a bruised face, and no hair, I probably wouldn't have anyway. We walked into the headquarters building, and the prisoner was marched in behind us. Ninh got up from his chair and began chewing the man out. Then he ordered the big sergeant to take the man back to the shed that was being used as a jail. Ninh walked over to us, shaking his head. "Lieutenant Vinh was a very bad man."

Jesus! The prisoner had been the 11th Company commander! We immediately asked what had happened. Gambling was always a problem in the battalion, and there were rules against it. No one saw anything inherently wrong with it, but if soldiers lost their money, they would have nothing to support their families with, so it was forbidden. Lieutenant Vinh had found four soldiers gambling, and in a rage, he had pulled out his .45-caliber pistol and began waving it around. Either because of his anger or by accident, the pistol went off and blew a hole in one of the gambler's legs. Unlike the movies, a gunshot anywhere in the body is almost always serious. This round had passed through the bone, taking out a couple of inches of the man's femur. He would recover, but one of his legs would be shorter than the other, and he would be an invalid for life. When Ninh heard what had happened, he hit the ceiling. He arrested Vinh, beat the hell out of him, shaved his head, and sentenced him to three days in jail. For the rest of our time at the training center, the big sergeant was to follow him around, monitoring his every move.

Vinh wasn't the only guy to get in trouble with a gun while we were at the training center. One night I was awakened by the unmistakable sound of bullets snapping over the top of the American officers' sleeping quarters. I rolled out of bed onto the floor and grabbed my rifle and web gear. There was a lot of yelling and crashing about as the other officers did the same thing. I crawled out to the walkway between the hootches and shot out the light on a lamp post overhead. I yelled for others to douse all lights and stay on the ground. I slithered down the walk until I reached the edge of the hootch. Then I carefully looked around the side of the building, expecting to see muzzle flashes and VC coming across the lawn. Everything was dark, except for the light in Likens' room, which dimly shined onto the porch. There he was, standing in the door, holding his carbine. I started to yell for him to get down when he fired off a burst over the top of the officers' hootches. It hadn't been the VC—it had been my sergeant!

I yelled for everyone to stay in place and hold their fire. Then I yelled at Likens, telling him that I was coming over to his hootch. He mumbled something that I couldn't make out and then turned and walked back into his room. Again I yelled for everyone to hold their fire while I checked out what was happening. Nobody argued, particularly since I was about the only officer in the compound with combat experience.

I got up and ran across the grass, onto the porch, and into Likens' room. He was sitting on his bed, drunk, with his carbine across his lap. I snatched it away from him, popped the magazine free, and ejected the round out of the chamber. Then I proceeded to chew his ass. As I was chewing on him, I was picking up the spent cartridges from the floor.

Although he was drunk, he soon realized what kind of trouble he had gotten himself into, and an "aw shit" look appeared on his face.

I told him to quickly get the brass (cartridges) policed up, and I added, "For God's sake, don't miss any." He was sobering up fast, and although he had a little problem with hand-eye coordination, he began to pick up the spent ammo. I spotted a cleaning rod in the corner and quickly ran a patch through the bore of the carbine and poured a gob of oil in the breech. I heard shouting back at the villa and someone yelled, asking what I had found. I yelled back that I was searching the area and that everybody should stay where they were. I picked up the brass that Likens was holding and made a quick check to see if we had it all. Then I dumped the water out of my canteen and dropped the cartridges into it. (All I was wearing was shorts and my web gear, so I didn't have any pockets that I could put them in.)

I told Likens to keep his mouth shut and when questioned, he should tell his interrogators that he had been asleep. If he said anything else, they would court-martial him, and he'd spend the rest of his tour in Long Binh Jail. I also said that if he told them what I had done, I'd end up in the next cell. Hoping that he fully understood, I went outside and gave the all-clear. A few people moved cautiously down the hill toward us, led by the full colonel. When he got there, he took one look around and figured that Likens was responsible. "So your man decided to shoot up the compound, did he?"

"No sir. I figure that it was a lone VC—or more probably, a drunken Ranger from down the road."

"Don't hand me that shit! This goddamn bastard had a couple too many and decided he would pull our chain. I'll have him court-martialed and sent to jail!"

"Sir, Sergeant Likens was asleep, just like the rest of us."

Turning to his lieutenant colonel, he said, "I want this area completely searched for brass. Take the bastard's carbine and bring it to my office." Then to me he said, "I want you and Sergeant Likens to report to my office, *immediately!*"

Hoping that Likens was sober enough to know what to say, I marched with him up to the colonel's office, escorted by the lieutenant colonel. We were forced to wait outside his door while his men searched the area. In ten or fifteen minutes a major went into the office, and I heard him report that they hadn't found anything. He was told to continue searching "every inch of that bastard's hootch —and every inch of ground all around it."

We waited another thirty minutes, and then the major came back and said that they had done exactly what the colonel had ordered but still hadn't found anything. I was hoping that Likens was sobering up, but I couldn't talk to him because the damned lieutenant colonel was watching both of us. Finally the colonel bellowed, "Bring in Likens and Parrish!"

He was sitting behind his desk; Likens' carbine was lying across it. Looking at me, he angrily snapped. "You're going to stand by this bastard?"

"Sir, I don't *know* that he did anything. When I got to his hootch, there was no sign that he had fired his weapon and no indication that he had done anything."

"You mean to tell me that he was just being a good soldier, sitting in his hootch and cleaning his weapon—drunk?"

"No, sir. He was lying on his bunk when I got there."

"You must think I'm an idiot! This worthless bastard didn't fire his carbine at the officers' quarters, pick up all the brass, and then clean his weapon before we got there. You helped him cover up. Shit! You did the cover up. He's not sober enough to have done it himself."

"Sir, I resent that. If I covered up for him by picking up all the brass, where the hell did I put it? I don't have any pockets on these skivvies [shorts]."

"I don't know, but you are to wait outside. I want to talk to Likens alone."

Hoping that the brass wouldn't jingle in my canteen, I saluted and carefully walked out the door. I waited another thirty minutes while the colonel browbeat Likens, trying to get him to admit his crime. I couldn't hear what Likens was saying, but since the colonel kept yelling at him, I figured that he wasn't confessing anything. Finally the colonel opened the door and told Robinson to come in. He'd been standing in the hall the whole time, but he hadn't said a thing.

A couple of minutes later, Robinson emerged and said that the colonel had ordered him to take Likens to Nha Trang the first thing in the morning and put him on a plane back to Division. When we got outside and away from the others, Robinson asked me in a low voice if Likens had really shot up the compound.

"Fuck no! He was drunk in his bed. It must have been one of those rangers from down the road."

Shortly after Likens left, Mosley packed up as well. His tour was over, and he was headed back to the States. They were both good men, and I regretted losing them. I hoped that Division would quickly send replacements so that I would have time to work with them before we returned to operational status. It helped a lot to know your people—and for them to know you—before you got into a firefight.

I don't know what the training center colonel had written to Colonel Sonstelie, our division senior advisor, but in a week he flew up to visit us. The regimental commander and Major Humphry came with him, and there was the usual dog and pony show briefing for them. I guess that I had pushed the training center colonel too much, because I wasn't invited to the meeting. It was okay, because no one ever told the whole truth at these briefings, only what they wanted the other people to hear.

Just before he left to go back to Lam Son, Colonel Sonstelie took me aside and told me that I was doing good work. He never mentioned anything about Likens or the thing with the tax collectors. He did say that he was sending us a new NCO, a man who had been with another unit for a while before he was wounded. The man had recovered, been released from the hospital, and would be joining us in a couple of days. The colonel said that we'd have to wait a little for a second sergeant. Then in a lower voice, he said, "But we'll get you another one, because we can't have just one officer and one NCO—Captain Robinson will be leaving you when you get back."

A few days later I drove to Nha Trang to pick up Likens' replacement, Staff Sergeant Wilson, who had flown up from Tan Son Nhut. He was in his mid-twenties, and for a man who had been in Vietnam four months, he had very white skin. He had lost his tan during a lengthy stay in the hospital. The personnel people had sent him to MACV because they felt that he deserved an "easy" time for the remainder of his tour. When I told him that he'd been assigned to the wrong battalion for an easy time, he was more than a little upset. "Every bastard I talked to said that I had lucked out being assigned to the 3d Battalion, because you guys were living it up in Nha Trang."

"No sweat, Sarge. We've got a few more days before we go back to the bush."

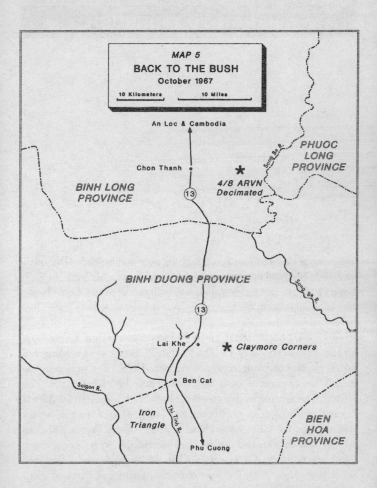

MAP 5
BACK TO THE BUSH
October 1967

10 Kilometers   10 Miles

An Loc & Cambodia

Chon Thanh ●

*
4/8 ARVN
Decimated

PHUOC
LONG
PROVINCE

Song Be R.

BINH LONG
PROVINCE

(13)

BINH DUONG PROVINCE

Song Be R.

(13)

Lai Khe ●

* Claymore Corners

Saigon R.

Ben Cat ●

Iron
Triangle

Thi Tinh R.

Phu Cuong

BIEN
HOA
PROVINCE

111

When we got back to Tan Son Nhut, we were met by Major Humphry and the regimental commander. There had been a big battle in Binh Long Province, and our battalion was being ordered to Chon Thanh, fifty kilometers north of Phu Cuong, about halfway to the Cambodian border.

At Chon Thanh, our CH-47 Chinooks landed on a very busy airfield next to Highway 13. Hueys and Chinooks were landing and taking off in a constant stream of aircraft. The Hueys were bringing in dead and wounded from where the battle had taken place. The wounded were being loaded on other Hueys for transport to the ARVN hospital, and the dead were being laid at the edge of the runway for loading into the Chinooks. There appeared to be over a 100 dead ARVN soldiers who had not yet been evacuated, with more arriving every few minutes. An American Catholic chaplain and a Buddhist priest were going down the rows of bodies, giving last rites and whatever rituals Buddhists had for their dead.

I was struck by the grimness of it all, particularly the heaps of bodies in the Hueys. It was obvious that everyone was hurrying to get the casualties back to Chon Thanh as quickly as possible. The dead soldiers had been piled in the back of the choppers like firewood—with heads, arms, and legs dangling out the doors. When a chopper landed, groups of men would untangle the bodies, haul them off the aircraft, and dump them on the ground. Then two-man teams would roll the bodies in ponchos and lay them at the end of a long row of dead. I noticed that besides their other wounds, most the soldiers had been shot in the head—it looked as if they had been executed.

Robinson and I were directed to the temporary command post in the compound on the other side of the airfield. A battery of six 105mm howitzers was firing continuously, adding to the noise of the helicopters.

The command post was teeming with people, including General Thuan, the division commander, and Colonel Sonstelie. This was obviously a big deal if General Thuan had left division headquarters and come out to the field. When Colonel Sonstelie spotted us, he motioned for Robinson and me to join them at the big wall map.

Earlier that morning the division's 4th Battalion, 8th Regiment had conducted an airmobile operation about ten kilometers east of Chon Thanh, near the Song Be River. One rifle company had gone in first and found the landing zone cold. The soldiers secured the LZ, and a short time later the battalion headquarters and a second rifle company landed. Before the slicks could bring in the third company, the troops on the ground spotted a couple of VC running away and charged off after them. They were following the bad guys up a small ridge when suddenly a hidden NVA [North Vietnamese Army] battalion opened up on them. In just a few minutes, they were surrounded by an additional NVA battalion and hit with 75mm recoilless rifles and 12.7mm (.51 caliber) heavy machine guns. The fighting was so intense that the third rifle company had to be inserted in another LZ about 800 meters away. Before anything could be done, the NVA overran the surrounded group and wiped out virtually everyone, including the battalion command group and three of the four advisors.

The 8th Regiment's commander asked Division for reinforcements, but none could get there in time, so the slicks had to go back and pull out the third rifle company before the NVA could get to its LZ and wipe it out, too. After killing all the ARVN and American wounded they could find and collecting the weapons, radios, and equipment, the NVA withdrew into the jungle.

The division's 5th Reconnaissance and Recondo* companies were now out beating the bush looking for the NVA regiment, and there were several aircraft flying around trying to spot Charlie. Colonel Sonstelie told us that our battalion was to stand by. If the NVA were located, we'd air-assault into the nearest open area. We only had enough Hueys to lift one company at a time, so we'd have to piecemeal in, just like the 4th Battalion. When Sonstelie was finished, I asked him how much chance he thought that one company would have against a whole NVA regiment.

"We believe that they have broken down into smaller units because of our search efforts. They know that if we can catch them all together, we can hit them with artillery and air and bring in a lot of infantry."

"Sir, when you say, 'smaller units,' do you mean smaller than a battalion? If not, we might lose another rifle company."

Knowing that I would be going in with the first company inserted, he put his hand on my shoulder and said, "We think so, but remember we've got a lot of artillery available, gunships standing by, and airstrikes on call."

I just nodded, because while I wasn't ready to get wiped out like the 4th Battalion had been, it was my job and I couldn't think of any better plan. I went back to the airfield and found that 10th Company had drawn the short straw, so I dropped my rucksack and sat down next to Lieutenant Tien. I didn't have to ask him what he thought, because it was clear from the expression on his face that he was worried, too.

By nightfall it was too late to insert us, even if our people did find Charlie, so we were told to stand down until the next morning. I picked up my rucksack and walked back to the compound. The command post was still busy as hell, so I went into the next room to get away from all the activity. The room was empty, except for an American NCO sitting in the corner. I said hello and dumped my equip-

---

*The term "Recondo" is a contraction of the words "recon[naissance]" and "[comman]do."

ment next to a wall. The sergeant mumbled a greeting and went back to staring into space. I pulled out a box of C-rations that I'd picked up in the other room and started eating dinner. The sergeant sat quietly for a while and then got up and came over. He looked like hell, and I could see tear streaks on his dirty face.

He turned out to be the only American survivor in the battalion. His captain had assigned one American to each company, and he had been with the one that had been extracted before the NVA got to it. He told me about listening to his buddies on the radio during the battle. At first they had been excited because they were chasing the VC. Then when they hit the ambush, the transmissions turned grim. The captain had been wounded and the lieutenant had been hit trying to get to him. The sergeant with the lieutenant took over the radio, and he told the captain that the young officer was in bad shape but that he was still alive. The captain radioed back that he had taken a couple of rounds in the legs but thought he'd be okay if they could hold out. They couldn't, and the last transmission was to the surviving sergeant. In a calm voice, the captain radioed that the NVA were killing the wounded—and they were moving toward him.

I asked the sergeant if he knew how many had been killed. Choking, he replied that he wasn't sure, but he thought it was more than 200. Besides his company, there were some who had survived by playing dead or who were so badly wounded that the NVA thought they were dead. A few had escaped by hiding in the thick bushes.

As I listened to the sergeant's vivid description of the battle and his friends' last moments, it occurred to me that my original orders had assigned me to *this battalion*. Division had two vacant assistant battalion advisor slots, and without realizing that I was already on orders to the 4th Battalion, the admin people had assigned another new lieutenant to my slot. When I showed up, they changed me to the other position with 3d Battalion. Had it not been for this minor paperwork screw-up, I would have been in one of the body bags that I'd seen being loaded into a chopper.

The next morning, our people had still not found the elusive NVA regiment, but they thought that it might have moved to the southeast, near the Song Be River, where the borders of Binh Long, Binh Duong, and Phuoc Long provinces met. Lieutenant Tien and I received orders to load the company into Hueys and air-assault into that area. Division was worried about losing another ARVN company, so we had to fly wide circles near the planned LZ while artillery and aircraft pounded it. Helicopter gunships blasted away with rockets and machine guns as we landed, and then they raced around above us, ready to fire again if the LZ turned out to be hot. It wasn't, but we sure alerted any NVA in the area to exactly where we had inserted. After an hour or two of checking around the LZ, the slicks came back in and picked us up. Off we flew to repeat the exercise at another LZ a few kilometers away.

I figured (and hoped) that the NVA had left the area. They would have been fools to stay. If they had been located, several ARVN and U.S. battalions, with massive air and artillery support, would have been on them. As I expected, the area around the second LZ was no more productive than the first had been. After we reported our initial findings, Division ordered us to begin moving to the west. The order was an admission that they had been wrong in their guess on which way the NVA had withdrawn, because the new direction led us toward Chon Thanh, about six kilometers away.

Finally Division admitted that the NVA had gotten away and that we were wasting our time, so we loaded into CH-47s and flew back to Lam Son. The day after we returned, Captain Robinson was given orders transferring him to Division as a staff advisor.

As required by army regulations, Captain Robinson showed me the efficiency report he had written on me. I was certain he was going to give me something less than a glowing report, and I was surprised to find that he had given me ninety-eight percent of the maximum score possible. I did chuckle when I read the section on personal qualities (dependability, enthusiasm, initiative, moral courage, etc.) —he had given me a maximum score in nineteen of the twenty areas. The only one that he gave me an "above average" in, rather than an "exemplary," was "Tact."

What Colonel Sonstelie had hinted came true: I was appointed the Battalion Senior Advisor when Robinson left. This was a little unusual because Division had captains waiting for a senior advisor vacancy, but I got the job because I was now one of the most experienced officers in the division. In combat, experience counts more than rank.

When the colonel told me of my new appointment, he said that he would try to get me a lieutenant assistant as soon as he could. I replied that I could get along nicely without a new lieutenant if he could give me a couple of NCOs. I needed one to fill the current vacancy and would like the other instead of the officer. He laughed heartily when I told him that lieutenants were more trouble than they were worth because they were too young and inexperienced.

A couple of days later Division called for me to come to Lam Son and pick up a new NCO. Sergeant First Class Morales was from Puerto Rico, and this was his first tour in Vietnam. He didn't stop talking during the entire ride back to Phu Van. Although he had no combat experience, he was likable and a refreshing change from the more subdued people we previously had on the battalion team. *Everything* was interesting, and *everything* was exciting, and he could *hardly wait* for his first firefight. I smiled and told him *that* would wear off pretty quickly.

Sergeant Wilson was happy to have company, since he felt a little uncomfortable with just an officer. Now that I was really in charge, I sat both of them down, and over a couple of beers, I explained how we were going to work. Despite having been in country for several months, Sergeant Wilson really had little combat experience. He'd been wounded only a short time after he got to the bush. Since Sergeant Morales had no experience at all, I decided I would keep both of them with me for a while and not send either of them out with a company. Their job was to pay attention to what I did and how I did it and learn what to do as quickly as they could. I wasn't going to throw either of them into combat alone, so if I really had to send someone, I'd send both of them. I would rather operate by myself than force either of them to be alone. They were pleased with that arrangement, but I warned them that as soon as we got the fourth man and I felt that they had a handle on things, we'd be operating in the normal two-man teams.

A couple of days later Ninh and I were called to division headquarters. The U.S. 1st Infantry Division wanted an ARVN battalion for a joint operation, and we were the logical choice. Our battalion and one from 1st Division were going to run a cordon and search operation east of Lai Khe, at a place the Americans called "Claymore

Corners." It seemed that every time an American unit went near the intersection of two dirt roads, Charlie would detonate a claymore at them. The Americans had lost quite a few people over time but were never able to kill any VC. The people firing the mines were local guerrillas, and everyone in the nearby village had identity papers showing that they were good citizens. However, one of the guerrillas had just *Chieu-Hoi*'d to the government's side and was willing to go with us to point out his former comrades. *

Our battalion was to be picked up by Chinooks and flown to a firebase a few kilometers from Claymore Corners, where we would be given a final briefing. Then during the night, both we and the other battalion would move north and surround the village. At dawn a National Police unit would come in with the turncoat, search the village, and arrest the guerrillas.

When the choppers landed us at the firebase the next afternoon, Ninh and I walked into the compound to meet the American battalion commander and his brigade commander (a full colonel). Ninh was a little nervous because he hadn't worked with American units before, and he asked me to "be in charge when we talk to them." I was a little self-conscious, because I was only a first lieutenant, so I made certain that my right shirt collar was under the shoulder harness of my web gear, concealing my black lieutenant's bar. I also made certain that my left collar, with the infantry crossed rifles, was showing and that my equipment didn't cover my Ranger tab, CIB, or parachutist badge.

The lieutenant colonel commanding the battalion was very friendly to both Ninh and me, but the brigade commander was one of those high-level, untouchable types who only condescended to shake our hands. The brigade operations officer briefed us on the routes to take to the village, the sectors of responsibility, radio call signs and frequencies, and the other details we would need to operate jointly. When he was finished, he asked if we had any questions. I asked the

---

*Under the South Vietnamese Government's *Chieu-Hoi* program, enemy soldiers were rewarded for defecting to the government's side. Defecters were properly known as *"Hoi-Chans,"* but American GIs routinely referred to them as *"Chieu-Hois."*

battalion commander how many troops he had, and he replied that he would be taking 250 on this operation. Then I turned to the brigade commander and said, "Captain Ninh has almost 350 soldiers . . . maybe he should be in charge."

Giving me a cold, hard look, the old bastard said, "Thank you, lieutenant, but I think not."

That night it was as dark as hell, and there weren't many landmarks to guide us, so we had to move most of the way to Claymore Corners by compass. I kept watching my compass and counting my paces to double-check the point team, because one of the most dangerous maneuvers in combat is the night movement of two friendly units in the same area.

About midnight, the word was passed back that we had reached the edge of the trees, a couple of hundred meters from the village. Whispering into the radio mike, I called the U.S. battalion and reported that we were at our release point, where the companies would fan out and move into their positions at the edge of the village. The U.S. battalion commander radioed back that his people still had some distance to go, and he didn't know exactly how long it would take them to reach their release point. This made me feel good, because we were *just ARVN*, while they were *well-trained Americans*.

The village was just south of the intersection and consisted of sixty or seventy hootches arranged in two rows, one on each side of the road. It was several hundred meters long, but only fifty to seventy-five meters wide. If we could get into position without being detected, we had enough troops to set up a tight perimeter all around the village, making it very difficult for any of the VC guerrillas to slip through our line.

The plan called for the two battalions to tie in at each end of the village; the road would serve as the boundary between them. The Americans were still unsure of where they were but thought that they were very close to the village. To help them, I agreed to move to the edge of the road at the southern end of the village and signal with a red-filtered flashlight. I flashed my light on and off several times and then saw a red dot in the distance, indicating that they had seen my signal and now knew exactly where they were.

Although the flashlight threw a directional beam and I had shaded it with my hand, someone in a nearby hootch had been alert and spotted it. The next thing I knew, a blast of AK-47 fire went snapping over my head. I threw myself on the ground and quickly crawled behind a small concrete fence post. Since the post was only about four inches in diameter, its protection was more psychological than physical. I heard another burst of AK-47 fire on the other side of the road, this one aimed at the Americans, and instantly, just about everyone in the U.S. battalion opened fire on the village. Hundreds of M-16 rifle and M-60 machine gun rounds were cracking through the thatched hootches and passing just above where my troops were lying. Ninh yelled at me to call the Americans on the radio and get them to stop their firing before we started taking casualties. Frantically I contacted the U.S. battalion commander and told him that his fire was coming right at us. He responded that his people had taken fire from the village and were returning it.

"I understand that, but your rounds are hitting my people. Request you cease fire *immediately*!"

"Roger. I'll get it turned off as quickly as I can."

I heard them yelling, and the shooting began tapering off, and then finally it stopped. I called to Ninh, asking if any of our people had been hit, but before he could answer, there was another burst of AK fire from the bastard who had originally shot at me. At the same time I was clawing the ground behind the fence post, the Americans opened up again.

"Goddamn it! Your people are shooting us again! If they don't stop, I'm going to give permission for my people to fire back!"

Before I could say anything else, I heard a "twang" and discovered that the antenna on the radio beside me was now twelve inches shorter. I groped around in the dark and pulled a long antenna from the bag attached to the radio's case. As quickly as I could, I unscrewed the bad antenna and replaced it with the good one. When I was done, I grabbed the handset and heard the American colonel calling me.

"I was off the air for a couple of minutes. One of your fucking rounds shot off my antenna! I'll give you thirty seconds to cease fire. If you don't, we're going to pour everything we have into your position!"

Not waiting for a reply, I threw down the handset and told Ninh

120

to get ready to tell our troops to open up. Until then, none of our people had fired, because unlike the Americans, they were afraid of hitting the friendlies on the other side of the road. It took the colonel just about the full thirty seconds to get his people to stop firing.

When it stopped, I got back on the radio and told him that while there were VC in the village, there weren't enough to justify the kind of firing his people had been doing. I told him I had some son of a bitch about fifty meters away who was taking periodic potshots at me *personally*, but I hadn't returned the fire for fear of hitting some of his people. I expected him to control his battalion until daylight, and then we'd go into the village and take care of Charlie. Until then, we had to agree to use only hand grenades, because their blasts wouldn't carry into each other's lines.

Several times during the night, a guerrilla would try to break through us, but each time he'd be killed. I hoped that my personal Charlie would try it, but he didn't. Instead, he fired at me three more times before it got light. I couldn't throw a hand grenade far enough to get him, so I just had to hunker down, hoping that he'd continue firing too high.

When the sun rose, Ninh sent two platoons into the village to round up all the peasants and herd them to the small schoolhouse near the intersection. Then we began thoroughly searching each hootch, looking for weapons and military equipment. I decided to personally search the hootch that Charlie had been in when he shot at me. I was tempted to burn it down, but because of the big stink the Marines had caused by doing it on television at Tam Ne in 1965, I decided that I'd better not risk it. I did find some AK-47 brass that Mr. Charlie had neglected to police up during the night, so when I was done searching the hootch, it needed a lot of repairs.

About the time I completed my search, a chopper brought in some National Police and the ex-VC, who was wearing a sandbag over his head, which had two eye holes cut in it. The villagers were lined up on the road, and one-by-one they were escorted to the schoolhouse to have their papers checked. When a policeman finished checking the peasant's identity card, the individual was directed to stand in front of the doorway. The hooded *Hoi-Chan*, looking like a medieval executioner, either nodded his head yes or shook it no. Whenever he nodded yes, the peasant was roughly shoved over to a growing group of people being guarded by some of our troops.

When the *Hoi-Chan* finished pointing out the VC guerrillas (or whoever may have pissed him off when he had lived in the village), we took the suspects back to their hootches and thoroughly searched them again, this time with less regard to their possessions. This second search netted several SKS carbines, an AK-47, and a couple of hundred rounds of ammunition.

While the police were doing their thing, squads from both battalions were searching the area around the village to see if they could find anything. One of my groups discovered a dead guerrilla; he had gotten through our lines but was so badly wounded that he only made it a couple of hundred meters before dying. The men carried him back to the village on a pole and set him down near the other dead VC. He stood out from the rest of the group because his body had stiffened into the position he was in when the troops had carried him in, with his arms and legs sticking into the air. In Ranger school we had called it the "dying cockroach" position.

We were done by midafternoon, and the Chinooks came in to take my battalion and the arrested suspects back to Lam Son. As I was breaking my people into helicopter loads and sending them to each CH-47, I noticed some words painted on the schoolhouse wall. In English, the VC had written; "American GI: Let the Vietnamese settle themselves their own problems. Refuse to fight and demand repatriation." It reminded me of the "Welcome to Phu Hoa Dong" painted on one of the buildings in the RF compound there—it had probably been written by the Viet Cong, too.

10 AN SON BASE CAMP

Not long after the Claymore Corners operation, Major Humphry came to us with some good news—we were going to get our own base camp and a permanent area of operation. At last some other battalion could get out in the bush and do what our battalion had been doing for almost a year—running from one end of the division's AO to the other.

I grabbed a map and had Humphry draw the boundaries of our new area and mark where the compound was located. He started drawing just south of Phu Van and traced a line down Highway 13. (Great! We'd be near civilization.) Then at the village of Bung, he drew the line southwest, to the Saigon River, and then northwest along the river. (*Shit! Nothing but mud and swamps!*) I had lost almost all my enthusiasm by the time he had completed drawing the irregularly shaped box. With a little dread, I waited for him to pinpoint the location of the compound. Just as I feared, he put a black mark in the southwestern part of the AO, about 500 meters from the Saigon River—as far from civilization as one could get and still be within the area. Oh well, at least we'd have a permanent home.

Wilson was really excited, even when I explained what a miserable area we'd been given. He couldn't have cared less, because he figured anything was better than running operations in places like the Iron Triangle, the Hook, and Phu Hoa Dong. Morales hadn't been in country long enough to have an opinion. I outlined the good and bad points for him, but the only thing he asked was, "Can I get my CIB there?" I assured him that he could and *would*.

When Humphry left, I went to find Ninh to see if he had gotten the word yet. He had and he was pleased. His officers had told him what a rotten place it was, but having a permanent compound so close to Saigon (about twenty kilometers away) meant that he could move his family from Nha Trang to the capital. Lieutenants Tien, Chung, and Vinh were much less happy about our new setup.

An Son, our new camp, was a disaster, and when I saw it, I was ready to desert. Except for the barbed wire and sandbags, the compound didn't look much like a military installation. There were bits of canvas, sheets of rusted tin, old ponchos, and polyethylene plastic throughout the camp to provide shade and some protection from the monsoon rains. Wherever there wasn't a bunker, there was mud and stagnant pools of water. The former occupants had dug down to the water table, just a few inches below the earth's surface, and had used the dirt to make small islands so they could build bunkers and weapon emplacements. Because the ground was so soft, the bunkers looked like poorly constructed log houses, with walls of crooked poles laid horizontally to keep the sandbags from falling over into the mud. Sections of perforated steel runway matting had been laid on sandbags to form raised walkways above the mud and water. Around most of the whole compound was a mud berm and a water-filled moat, where the earth had been taken for the fortifications.

In the center of the place was a big tent surrounded by four-foot-high sandbagged walls. It was the largest structure in the camp, and as it turned out, it was where we advisors were to live. Inside the tent our predecessors had laid a floor using sheets of plywood on top of sandbag foundations. It wasn't very stable, and it felt like a trampoline when we walked across it. At one end of the tent were four of the most unusual beds that I'd ever seen. Over each canvas cot they had built a contraption of wood and sandbags. At the corners of the bunk they had placed vertical wooden four-by-fours. On top of the posts was a sheet of plywood covered by a layer of sandbags. Because they couldn't build overhead cover for the whole place, they had built these contraptions for each bed. I didn't know what they had expected to accomplish, because it was obvious that the covers wouldn't provide protection from anything. The weight of the sandbags caused the wooden canopies to sag so much that if anything larger than a rock hit them, they would collapse on the beds. Wilson, Morales, and I just stood there for several moments, wondering what we had done to deserve this.

I didn't know about my teammates, but I didn't sleep very well that night. No matter how well I tucked in the mosquito netting, the little bastards kept finding ways to get through it and bite me. Shortly after dawn I awoke and decided to find a place to relieve myself. I swung my legs over the side of the bed and found that I was standing in ankle-deep water. Although we were more than sixty kilometers from the South China Sea, the land between An Son and the coast was flat. Whenever the tide came in, the Saigon River backed up and flooded the low ground. Twice a day we would watch the water slowly flow into the compound and rise to the foot of the bunkers. At high tide, about four inches of water covered the floor of our tent—and anything else we had been foolish enough to leave lying around. It was an altogether miserable place to live.

After a few days of moving our stuff and trying to get our tent organized, I took the jeep and drove to the division's advisor compound at Lam Son. I wanted to get a new mosquito net before I was drained of so much blood that I'd have to be evacuated. As I was talking to the supply sergeant, I noticed several large cardboard boxes with the word "Television" printed on them. I asked about them, and the sergeant said that they were being issued to battalion and regimental advisor teams that were in static locations. I reminded him that we had just moved into An Son, and so I walked out a short time later with a brand new, 21-inch, black-and-white Panasonic TV.

When we got it unboxed and had it sitting on a table (well above the water), it occurred to me that we didn't have any electricity to run the damned thing. "No sweat," said Morales. "I'll have us a generator within two days. All I need is a VC flag and one of the AK-47s we've got stashed for trading."

Sure enough, two days later a Huey landed on the pad outside the berm, and there was a grinning Morales with a 1.5-kilowatt generator. He told us that he had driven around Long Binh until he found a fellow Puerto Rican, who knew another Puerto Rican who would trade him a generator for the assault rifle and flag. According to Morales, Puerto Ricans stuck together, and he could get almost anything we needed once he made connections with what he called the "Puerto Rican Mafia."

For the first couple of days, the TV was great, but it soon became a real pain in the ass. Every evening, Ninh and the officers brought their chairs, and a hundred or more soldiers would crowd around the tent to watch it. We had absolutely no peace and no privacy until late at night. When the Vietnamese became bored with our programs, they began to ask that we turn to their channel so they could watch amateur imitations of American shows and drama heavily laced with pro-government propaganda. I regretted getting the damned thing and began figuring out how I could get rid of it.

On the seventh night of this bullshit, I was lying on my bunk and trying to write a letter while the TV blared Vietnamese music at full volume. I had just about decided that I would give the Vietnamese the damned thing—with the proviso that they would have to use it on the other side of the compound—when *Boom*! Something exploded about halfway between our tent and the berm, raining mud down on the compound. I yanked the plug on the TV, and everyone ran for the bunkers. A few seconds later there was another explosion, very near where the first had been. Wilson, Morales, and I raced out of the now blackened tent and into the command bunker next door. Ninh was on the radio, calling for reports from the outposts and ordering the company commanders to be ready for a possible attack. After the initial noise and confusion, everything was quiet, except for the sound of our 81mm mortars firing every few seconds and the pop of the illumination rounds overhead. We remained on alert for the next two hours, but nothing further happened.

Early the next morning I searched the area and found two craters where 60mm mortar rounds had impacted and exploded. Because the tide had been in, the only real damage was a few shrapnel holes in a couple of tarpaulins and some torn sandbags. That afternoon I had Wilson and Morales help me carry the TV to the jeep so I could turn it in to the supply room at Lam Son. When I got back, I told them that we were going to build our sandbagged walls much higher than the four feet that our predecessors had felt was sufficient.

Having gotten rid of the television, I decided that we still needed a little entertainment, so I had Louie, our Vietnamese driver, take me into Saigon. Near the big American PX in Cholon, I spotted what I was looking for: a Vietnamese standing beside the street with a monkey. I had Louie stop next to him, and I began to barter. After five minutes of haggling, we settled on a price of three cartons of Salem cigarettes.

I bought the cigarettes in the PX and then had Louie take me back to the monkey man. We drove down the street and slowed in front of the guy. Without stopping, I grabbed the leash around the animal's neck and thrust the paper bag containing the cigarettes in the man's hands. Without waiting for him to check to see if I was ripping him off, I told Louie to step on it. We roared down the street, looking over our shoulders to see if any police were following to catch the "big-time" black marketeer. When I was certain that we were okay, I told Louie to slow down and take us back to An Son.

I don't think the monkey had ever been in an open vehicle. He was terrified and kept jumping out of my lap and grabbing Louie by the hair. We almost got into three accidents before I could get control of him. All the way back to An Son, he held on to my fatigue shirt for all he was worth, watching the scenery whiz by the jeep. He was a small rhesus-type monkey, with pointed ears and teeth like a human—in fact, he looked just like a little old man. (I found out about his teeth when he bit me several times during the ride home.) Louie asked me what I was going to name him, and for no reason at all I decided that I'd call him George.

Louie dropped me off at the bridge, and I walked into the compound with George sitting on my shoulder. Several soldiers spotted him and started calling, *"Con khi! Con khi!"* [Monkey! Monkey!]

Soon a small crowd had gathered around, poking and laughing at

George. Then one of them asked, *"Trung-uy an thit khi, khong?"* [The Lieutenant is going to eat the monkey?]

In very strong Vietnamese, I told them that I was *not* going to eat the monkey—and that *no one else better try to eat him.* I made a cutting motion across my neck to suggest what I would do to anyone who decided to make a meal of my new pet.

Morales and Wilson were lounging in chairs when I entered the tent. As soon as they saw the monkey, they jumped up and tried to pet it, but George bit both of them. It took several days and quite a few bites before George calmed down enough to allow us to hold him without struggling. When he did, we found out that George should have been named Georgia, but by that time it was too late, because we had gotten used to the name.

George was the best thing I had gotten in Vietnam since I had scrounged my M-16. She wore a small dog collar, and we kept her tied on a long length of field telephone wire. It was long enough so she could roam but short enough to keep her from destroying our stuff. Some soldiers had brought their families to live in the compound, and every day a pack of children would sit along the walls of the tent and laugh at George's antics. Captain Ninh thought she was funny until he tried to pick her up and she chomped down on his finger.

By the end of the first week, George had decided that I must be her father, because she would sit on my lap and pick the hair on my arms, looking for lice and bits of dirt. From what I had read, I knew that this was what monkeys did to each other to show affection. Whenever I walked into the tent, George would scream and race over to me. One night she got loose, and I woke to find her clawing at my mosquito net to get in bed with me. After that, whenever I spent the night in the camp, she slept curled up next to my head.

George was not the only female in my life. Not long after we had settled into An Son, Captain Ninh invited me to visit his family in Saigon. They had just moved from Nha Trang, and he was anxious to see them. I put on my cleanest jungle fatigues and best beret, and we drove into the city. After having lunch at their new house, which was just as nice as the one they had in Nha Trang, we drove to a park in the center of Saigon. We spent the afternoon strolling around, looking at statues and a very ornate Buddhist temple. The eldest girl was quite relaxed, and we carried on with small talk about the beauties of Saigon (there were few) and the country (there were many). The littlest girl stayed as close to her father as she could. Although Madame Ninh kept calling for the middle daughter to walk with her, the girl kept easing back to her sister and me, giggling all the time.

The next afternoon Ninh came to our tent and invited me to go up the road to Bung and have some Chinese soup. After eating, he ordered a couple of beers, and we sat back talking about Saigon and Nha Trang. Then he changed the subject and asked me if I liked his daughter. Concealing my uneasiness, I again told him that she was charming, intelligent, and pretty. He smiled and hit me with a bombshell. "Lieutenant Bob, would you like to marry her?"

I choked and sprayed beer all over the table. Stalling for time, I pulled out a handkerchief and began wiping the table dry. Then I apologized for my poor Vietnamese and said that I hadn't understood what he had said. He repeated his question and added that he and Madame Ninh would be very pleased to have me as their son-in-law.

"But Captain Ninh, I'm already married and have a baby."

"Yes, I know, but in Vietnam it is possible to have more than one wife. It is an old custom of ours."

"But I am an American, and our law forbids us from having more than one wife."

"Yes, you are an American, but you are *almost* Vietnamese, and you are in *Vietnam* now."

"What would happen when I have to return to the States? I couldn't take your daughter with me."

"No problem. You will come back after a few months, won't you?"

"Well, yes. If I'm not killed or badly wounded this time, my army will send me back for another tour in about a year."

"Okay, no problem then."

"Captain Ninh, you have a beautiful daughter that any man would be glad to have for his wife, but in my country it would be against the law. We also have an old custom in America—*a wife may shoot her husband if he messes around with another woman.*"

We argued back and forth for the next twenty minutes, with neither of us giving ground. Ninh became a little agitated but didn't get angry.

Finally he said that I should think about it. I did—and I always made certain that I had something that I *absolutely* had to do whenever he invited me to visit Saigon. He apparently got the word, because although he would drop comments from time to time, he never again pushed me to marry his daughter.

After the first week or so of getting the camp organized, the battalion had established an operating routine. Each day one company would sweep a part of our twelve-square-kilometer AO. Another company would send out squad- and platoon-sized night ambushes. The third company would remain in the compound to defend it against a possible attack and to be on call in the event the sweep company needed reinforcing. Every day the companies rotated duties.

At first I decided to go with the sweep companies to get a feel for the area. I took either Wilson or Morales with me so I could bring them up to speed. The other NCO remained in camp by the radio to get us support if we needed it. These daily operations were much easier than the ones that we had been running, because we didn't have to take our heavy rucksacks. On the other hand, they were more frustrating because the only VC permanently in the area were guerrillas. Occasionally a VC platoon or company would pass through the AO, usually at night, but most of the time we had to deal with one or two individuals who would snipe at us and then disappear. We averaged a couple of our soldiers killed or wounded each week because of these sniping incidents or booby traps. We were seldom able to kill any of the guerrillas because they could hide their weapons and blend into the population so easily. It was frustrating, but when we did kill one, we considered it the equivalent of killing at least ten main-force Viet Cong.

Once I felt that I had a good handle on what was in our AO, I decided that it was time to switch to the ambush company. Neither Wilson nor Morales liked that idea at all, and they reminded me that it was against MACV rules to go out with anything less than a company. I agreed and said that I wouldn't make either of them go with me but that I wanted one of them monitoring the radio during the entire night, just in case I got in trouble. I worked out a rudimentary radio code that we would use. They were to call me each hour by pressing the "push-to-talk" button on the handset three times. I would reply the same way to indicate that I was okay. Other codes included one click for "yes," two clicks for "no," and a long series of rapid clicks that meant I had VC near my position.

Ninh didn't like the idea of my going on squad ambushes, and he tried to talk me out of it. I told him not to worry; if he needed U.S. artillery, air, or medevac, either Wilson or Morales could get it for him. He frowned and said that he wasn't worried about getting support, he was worried about me getting killed. I laughed and reminded him that it was the nature of our business. Lieutenant Phuc, the battalion operations officer, and Captain Phat told me that I was crazy and declined my invitation to go along.

The first squad I joined for an ambush almost fell over from shock. The troops had never seen an advisor work with a single squad—and definitely never expected to see one go out on a night patrol. When the Vietnamese squad leader finally understood that I was serious, he gathered his seven men together and told them to make certain that nothing happened to me. I stopped him and told them that they should just consider me to be another patrol member—but that I would appreciate it if they could bring me back if I got wounded.

We left the camp shortly after dark, hoping that the VC wouldn't see us departing. Once we were in the woods, the patrol leader made several changes in direction so that if we had been observed, Charlie would have a hard time figuring out which way we had gone. We finally set up along a trail, at the edge of a small clearing. Everyone except me had brought a claymore, and we put some of them along "the killing zone." Additional claymores were positioned at each end of the zone, facing up and down the trail, in case there were more VC than the killing zone could handle. Two more claymores were put out behind us to cover our rear. Except for my M-16 and a BAR (Browning Automatic Rifle), everyone else was armed with semi-automatic M-1 rifles, and each of us carried at least four M-26 hand grenades.

It was very dark until about 2300, when a half-moon rose to cast an eerie glow. The trail to our left was still in darkness because the trees and overhanging brush blocked the moonlight, but to our right we could see the trail passing through the clearing and disappearing into the woods on the far side. After hours of lying perfectly still, I was stiff and beginning to wonder what the hell I was doing there. I really wasn't going to make any real contribution to the success of the ambush, and if Division found out what I was doing, I'd get my ass chewed. The mosquitoes were still after me, since I hadn't put on insect repellent (it could be smelled several meters away), and the little mosquito net that I had put over my neck and head didn't stop them; it only slowed them down.

Despite the bugs and my stiffness, I was just beginning to nod off when I felt a hand on my left arm. There were VC coming down the trail. A few seconds later two guerrillas walked less than ten meters in front of me. One was carrying an AK-47, and the other looked like he had an SKS carbine.

I waited for the squad leader to detonate a claymore, the signal to open fire. After what seemed like a very long time, the claymore on the right exploded. The squad leader had waited as long as he could to see if there were any other VC in the group. At the last possible moment he blasted the claymore, and everybody opened up.

After about a thirty second barrage, the squad leader gave the signal to cease fire, and three soldiers got up and moved to where the VC bodies were lying. Two provided security while the third searched the dead. I got to my feet and carefully walked over to the bodies. There was enough moonlight to see that there had been no need to fire our rifles—the claymore had blasted the two VC into bloody pulp. We quickly stripped the bodies of everything except their pajama shirts and pants and returned to the others.

The squad leader gave orders to pull in the other claymores and to gather up our grenades and the extra ammunition we'd laid on the ground next to us. When everything was accounted for, we pulled back from the trail and moved several hundred meters away from the site. We found some thick brush and set up a small perimeter to wait for dawn, when we could return to the camp.

When we got home that morning, everyone was waiting for us. There hadn't been a successful ambush since we'd been in An Son, and it was a big event. Ninh inspected the stuff we had taken from the dead VC and congratulated each member of the squad. He walked me back to my tent and asked that I not go out again, saying that he had been awake all night worrying about me. I thanked him for his concern but said that I would be going out again the next night.

Unfortunately, while we had several more successful ambushes, they were never the ones that I was with. Every time I heard an ambush go off in the distance, I cursed my lack of luck for having chosen the wrong patrol.

I'd gone on four ambushes when I got a radio call asking me to come to division headquarters. When Colonel Sonstelie saw me in his outer office, he waved for me to come in and sit down. After some small talk he said that his people had been having a hard time contacting me at night. It seemed that I was always "checking the perimeter" when they called and that I never called back until after daylight.

Sensing that he might be thinking that I was screwing off and spending time in Saigon or somewhere else, I began to explain that with only three people, I had to devote extra time to checking our defenses. Sometimes my sergeants had forgotten to tell me about the messages until it was too late to contact anyone other than the TOC duty officer. Giving me a fatherly look, Colonel Sonstelie interrupted my lame explanation by saying that he knew that I was doing my duty—but he wouldn't like to lose an officer on an ambush that the officer shouldn't have been on in the first place. No ass chewing—just a simple statement.

Changing the subject, Sonstelie reminded me that I was just about to complete six months in the field, and he asked me if I had given any thought to what I wanted to do when I left the battalion. When I asked him about the sort of job I would get if I came out of the bush, he said he'd find something for me on staff or possibly as the advisor detachment commander.

"You *really* don't have many slots for a first lieutenant, do you sir?"

"No problem. You have been holding a captain's slot for the past several months, and we'll just put you in one here."

"Would you agree that I've done pretty good work for you since I have been in the division?"

"Certainly. I think that's obvious since you're about the only lieutenant in a battalion senior advisor's position."

"Well, sir, if I've satisfied my field requirements and you don't have a real need for me here, I would appreciate a transfer to a Ranger battalion. If I can't get that, I'd like to stay with my infantry battalion."

"You *want* to stay in the field? Don't you realize that you've been lucky so far and that your luck can't hold forever?"

"Yes sir, but I can be a staff officer any time in the States. This may be my only chance for combat duty. I'd really like a U.S. rifle company, but I know that MACV will never let me go. So I have to make the best of this tour—and that's in the bush, not in some REMF [Rear Area Mother Fucker] job."

Colonel Sonstelie smiled and said that he would see what he could do. We shook hands, and as I walked out of his office, he gave me a fatherly pat on the shoulder.

Just after midnight that evening, I heard our mortar fire an illumination round. I rolled out of the bunk, grabbed my M-16 and web gear, and ducked into the command-post bunker. Ninh was already on the radio, trying to find out what was going on. He was talking to Lieutenant Vinh, asking questions and listening to the answers. Each time he talked, he got angrier, until he finally threw the handset down with a hearty *"Du Ma!"* [Motherfucker!]

I waited a few moments for him to calm down and then asked what was going on.

"The northern outpost is reporting that they have ghosts throwing stones at them."

Not certain that I understood him correctly, I asked if he had said "ghosts," and he replied, "Yes—ghosts."

I suppose that most Americans in Vietnam would find this a little incredible, but it really wasn't. Besides belonging to a recognized religion, most Vietnamese are also animists. They believe that natural phenomena (trees, wind, rocks, thunder, etc.) possess souls and that the world is filled with supernatural spirits and demons, some benevolent and others wicked. The most common spirits are the *ma* [ghosts], which range from being just capricious to being very malevolent. Some *ma*, such as the ghosts of dogs or cats, are fond of playing tricks, like tripping people after dark. Other of the *ma* are much nastier, such as the *Ma Tham Vong* [Whispering Death Ghost] who tries to get people to commit suicide.

When the An Son camp was constructed, the soldiers built small sandbagged bunkers outside the perimeter as outposts to give early warning in case of a VC attack. These outposts were positioned about fifty meters forward of the earthen berm, in the middle of the mass of barbed and concertina wire that surrounded the compound. Every evening squads would climb over the berm, follow zigzag paths through the wire, and occupy all but the northern bunker. On the north side, the squad would set up around the bunker rather than in it. When I asked why the soldiers exposed themselves to possible VC fire, I was told that many of the troops believed that the bunker was haunted. It seems that sometime before we took over the camp, a soldier had committed suicide in the bunker by blowing himself up with a hand grenade. Since then no one had been willing even to enter the bunker, let alone occupy it all night long.

The next evening I decided to spend the night with the squad being detailed to the northern outpost. I was interested in the ghost business, and I wanted to show my Vietnamese that their advisors could handle demons as well as Viet Cong. Ninh told me not to waste my time, but other Vietnamese warned me "not to fuck with ghosts." I was careful not to make fun of their culture, but I couldn't resist commenting that I was only going after "VC ghosts."

At 1800, I met the squad at the berm and followed the men through the wire to the bunker. Besides my M-16, I carried an M-79 grenade launcher and a radio so I could talk to my team inside the compound. After I got to the outpost, Wilson called me on the radio for a communications check, using the call sign "Ghost Patrol Six."

I dropped my equipment and walked in the door of the bunker. *Christ, did it stink!* Holding my breath, I shined my flashlight around and saw that many of the sandbags had been ripped open from the blast of the grenade. The walls were splattered with black stains that appeared to be the remains of blood and flesh. The smell confirmed that they hadn't cleaned up all the mess and that the suicide hadn't happened much before our battalion took over An Son. Almost puking, I backed out the door and turned my head into the slight breeze to get some fresh air. I didn't know about ghosts, but I did know that there were very powerful reasons why no one would want to stay in the bunker. After I caught my breath, I noticed that my seven soldiers were staring at me with big eyes. I laughed, held my nose, and said, "Number Ten!"

Fortunately, the tidal water didn't come up to the north side of the compound, so we were able to spread our ponchos on the ground and lay on them. I chatted with the soldiers until it got dark and then pulled my poncho over me and went to sleep.

"Lieutenant, Lieutenant! Wake up!"

I grabbed my rifle and rolled over on my stomach, facing the shadow of the tree line in the distance. I looked to my left and right and found the soldiers doing the same thing. Whispering to the squad leader, I asked him what was happening. Even in the dark I could tell that he was shivering as he said, "*Ma!*"

I lay there for two or three minutes waiting for the "ghost" to appear, but I couldn't see anything. Then a dirt clod came sailing out of the black and hit the ground beside me. At first I thought that it was a grenade, but when it hit the ground, it splattered dirt in all directions. Shit! Wilson or Morales were throwing things to pull my chain. I grabbed the radio handset and called them. Wilson answered, and I told him to knock off the rock throwing. With surprise in his voice, he asked me what I was talking about.

"You know what I'm talking about. One of you bastards is throwing rocks from the berm. Cut that crap out before you panic my little people [Vietnamese soldiers]."

"Negative on the rock throwing. We're both in the tent, and we ain't throwing rocks."

"I don't believe you. Put Morales on the radio."

"Roger—wait one."

In a few seconds Morales called. If they both were in the tent, neither of them could be the rock thrower—but maybe they had taken a radio out to the berm. I told them to stay in the tent, and then using the squad leader's radio, I called Ninh and asked him to have someone check our tent to see if both of my people were there. In a couple of seconds Ninh called back on my radio and said that they were standing right there.

"Okay, so you guys got back to the tent before I could catch you."

"Ghost Six, we ain't got the foggiest idea of what you're talking about."

As I was telling them that it was a good joke but that now it was time to come clean, another dirt clod just missed my head. From the way it splattered back toward the berm, it was obvious that it had been thrown from somewhere out in the barbed wire and mine field, not from the compound.

"You're not going to believe this, but some silly-ass VC is throwing rocks at my outpost. Get the mortars to fire some illumination."

In about twenty seconds there was a *whoomp* and then a pop, and a flare began floating down on a small parachute. I scanned the barbed wire but couldn't see anything that looked like a human form. As the flare burned out, the mortar fired another one, but I still couldn't find anything. I looked behind me and saw a row of heads along the berm. They too were trying to spot the guy with the rocks—or the ghost.

After alerting the compound, I fired my M-79, walking the rounds all along the wire in front of me, but I didn't see or hear anything that would indicate there might be a man out there. After thirty minutes we stopped the mortar and everything grew silent. Whoever it was, he was either dead, wounded, or had pulled out. I rolled up in my poncho and went back to sleep, fully expecting to have dreams about ghosts and goblins.

Less than an hour later, another dirt clod hit my legs—the bastard was still there! This time I decided to let it go. If the son of a bitch had really wanted to kill us, he wouldn't have thrown rocks, and I wasn't going to give him the satisfaction of seeing me react.

The next morning I took my NCOs out to the bunker and showed them the bits of broken clod. I didn't want them to think that I had gotten spooked the night before. We searched along the northern edge of the barbed wire but couldn't find any traces of the visitor. I picked up a couple of dirt clods and heaved them as far as I could, but I couldn't come anywhere near the bunker. Whoever did it had to have crawled well into the tangled wire to have been able to hit the outpost. Just in case, I also tried throwing a rock from the berm, but it too fell short. All of us agreed the guy had been in the middle of the wire when he threw the clods.

Puzzled, Morales asked why someone would do something like that. I told him that it made sense if you knew something about Vietnamese culture. It was psychological warfare. If the Viet Cong could make our soldiers believe that An Son was haunted, it would have a much greater effect on our morale and nerves than just harassing us with weapons. *

A couple of days later I was cleaning my M-16 when Wilson returned from Lam Son, where he'd picked up our mail. With an odd look he announced, "I just met your replacement. He was processing in at admin when I dropped by to see if we had any distribution."

"No shit? I've got a replacement?"

"Yeah, he's a new dude, fresh from the States."

"What's he look like? You think that he'll be worth a damn?"

"No sir. I think that we're going to have problems. Is there any way you can keep this team?"

"It'll depend on what they've got in mind for me. If it's a REMF job, I'll raise hell, and I think Colonel Sonstelie will let me stay."

As soon as the road was clear the following morning, I was in a jeep headed for Lam Son. When I got there, I went straight to Colonel Sonstelie's office. I walked in, saluted, and told him that I had heard that my replacement was on board. Then I asked him what was going to happen to me. He smiled and said that MACV said that I could have a Ranger battalion if I really wanted it, but I'd have to start off as an assistant battalion advisor again.

"You mean I would have to work for some captain that has less bush time than I have?"

"I'm afraid so, but I have another assignment that I would like you to take; one I think you'll like."

"Okay, sir. What is it?"

"How would you like to be the senior advisor to the division's Reconnaissance and Recondo companies?"

---

* About a month later, after I had left the battalion, I ran into Wilson. He told me that the rock throwing happened a couple of more times—and then a VC crawled into the wire with a white sheet over his head. He was making ghostlike moans when one of the battalion's nonbelievers shot a hole in his chest. Since then, the unit hadn't had any more problems with Mr. Ma Tham Vong.

Captain Ninh had already found out about my transfer even before I got back to An Son. As soon as he confirmed it with me, he was in a jeep and on his way to see General Thuan. I was sad to be leaving the battalion after all we had been through together, but I wanted the 5th Recon Company. I hoped that Ninh wouldn't be able to convince the general to ask Colonel Sonstelie to revoke my orders.

I had to wait almost four hours for Ninh's return, but when he walked into the compound, I could see from the expression on his face that I could start packing. Ninh came into our tent and flopped down in a chair. He said that he had tried everything, including offering General Thuan a bribe, but the general wouldn't *an tien* [take a bribe; literally, "to eat money"]. I was really impressed, not so much because Ninh had offered money to keep me, but because Thuan had refused. (I needn't have worried, because the 5th Recon and G-2 Recondos were favorites of General Thuan, and he wanted the best advisors that he could get for them. Ninh's pleading had merely reinforced my selection.)

Ninh and his people threw the finest going-away party that I had ever attended, but there were many long faces. Combat brings men closer together and forges stronger bonds than any other experience. I was really sorry to leave the battalion, but I was also looking forward to joining a force that everyone considered the elite of the division.

When I got to Gosney Compound, where the division advisors lived, the billeting officer assigned me to the hootch I could use when I wasn't in the field. I didn't have to track down the man I was replacing; he was in the same hootch, packing to go home.

Captain Gene Bolin and I popped a couple of beers, and he proceeded to fill me in on my new units. He handed me a manila folder filled with newspaper and magazine clippings, all about the two companies. They'd even been written up in *Time* and *Newsweek*! One typical article, which was in the U.S. 1st Infantry Division's *American Traveler*, reported on a battle they had been in only a month before. The reporter had been very impressed:

> I felt uneasy when we lifted off, but not because of the Vietnamese outfit [the 5th Recon Company]—they were armed to the teeth. After the choppers dropped the Recon Company at the south end of the mortar and rocket pock-marked Loc Ninh, the Vietnamese solders spread out and moved toward the VC occupied sector. The Recon engaged the Viet Cong decisively. It was hard to keep track of them—they were everywhere. Some of them ran into the village, literally covered with weapons. I saw one soldier carrying an M-60 machine gun, two light antitank weapons, and a .45 caliber pistol. He fired the machine gun from his hip into a barn where several VC were trying to get a recoilless rifle into operation. . . . Summing up his battle experience with the elite ARVN unit, Waite [an American GI] said, "I wouldn't hesitate a minute to go in with them again. When I see a Recon scarf on a Vietnamese soldier, I know that I'll be with the kind of unit I want to work with."

The article was several pages long and included photos of Recon soldiers attacking through the village. I thumbed through the rest of the stack but couldn't read them while Gene was talking.

The 5th Recon and the G-2 Recondo companies were about as different from one another as was possible. The Recon was a well-

3d ARVN Infantry Battalion command group relaxing between firefights (left to right): Captain Kinsey, Major Man (battalion commander), Captain Phat (executive officer), and Lieutenant Phuc (operations officer).

ARVN discipline. Vietnamese soldiers being punished for minor offenses. A sentence of one or two weeks on bread and water was typical. Serious crimes were punished by a year or more in a frontline penal unit or death by firing squad.

Major Man in a pose he knew many Americans would believe
was typical of the Vietnamese Army. This photo was taken
shortly before the regimental commander fired him.

The author's farewell party before he leaves to take over the Reconnaissance Company. Captain Ninh is standing at the head of the table. The rifle company commanders, Lieutenants Tien, Chung, and Vinh are seated at lower right. (Yes, they served duck blood.)

Briefed and ready to take off for the Iron Triangle. From left to right are: Major Tu (division intelligence officer), author, Lieutenant Callan, Captain Hung (Recon Company commander), and Lieutenant Dac (company executive officer). An hour later, the Company was in a major firefight.

The author and Lieutenant Callan with a Recondo patrol, in front of a destroyed hootch. Many of the Recondos were criminals recruited from prison.

George, just before her first firefight. A few hours after photo was taken, George learned that she *didn't* like combat.

"You'd better be able to swim." A helicopter assault south of the Iron Triangle. The chopper is hovering, but the soldier is swimming for his life.

The Hook. Vietnamese and American troops spent the next twenty-four hours pinned down in the mud and water—it wasn't a happy Fourth of July.

House-to-house combat. Attacking into the North Vietnamese Army's 1st Battalion, 33d NVA Regiment in Phu Hoa Dong on February 21, 1968. In the confusion, the machine gunner and two riflemen inadvertently exposed themselves.

The warning came too late, and moments later the three soldiers were killed. A fourth, pointing toward the NVA machine-gun nest, was badly wounded. (The photographer was behind a large pile of logs and escaped being hit.)

Airstrike. F-4 Phantoms soften up the North Vietnamese in Phu Hoa Dong before the Recon Company resumes its attack. The NVA were well dug in, and it took hand-to-hand combat to root them out.

The tile factory where the NVA machine-gun squad was hiding. The author's antitank round blew the hole in the roof and killed the squad— although it wasn't where he'd aimed.

Body count. One of the NVA soldiers who ambushed the Recon Company's command group.

Ambush! The Viet Cong ambushing the Recon Company and the 1st ARVN Armored Cavalry Squadron on Highway 13. The lead tank is firing through the graveyard while the other tanks attempt to move up. The Recon has dismounted and is maneuvering to counterattack the VC.

disciplined force composed primarily of Nungs. The Nungs were a tribe of ethnic Chinese who had lived in Vietnam for hundreds of years. They considered themselves to be a warrior race, a people superior to all other groups in Indochina (Vietnamese, Cambodians, Montagnards, etc.), particularly at fighting.* All the company's officers and some of its NCOs were Vietnamese, but most of the troops were Nungs. Everyone in the 5th Recon was a volunteer, including the advisors. No American was acceptable until he had served six months in the field with a battalion and had proved himself in combat.

Gene warned me that I had better love to fight, because if I didn't, I wouldn't last long. The Recon would demand that I be replaced with an advisor who did. I had also better be prepared to be right up front, because that's where the company's leaders were when the shit hit the fan. They led the troops—they didn't direct them from several hundred meters to the rear. Looking me in the eyes, he said that no advisor had ever made it through a tour with the 5th Recon without being killed or wounded. Then he showed me a bandaged wound that he had gotten only a short time before. "I can show you some more, but I think you get the idea."

The 5th Recon was a hybrid unit. Every infantry division had a recon company, but this one was different. Originally it had performed the same type of missions that one would expect a reconnaissance company to conduct: long-range patrolling, small hit-and-run operations, and so forth. Then the division got a new G-2 intelligence officer, Major Tu. Tu was a sharp officer who was in tight with President Thieu and had a lot of pull. Tu had attended the U.S. Army's officer basic and advanced courses and its Command and General Staff College. He had been in the combat and intelligence (*not* "combat intelligence") business for a long time. I was told that he had even led a couple of teams into North Vietnam on sabotage operations.

---

*The Vietnamese looked down on almost all other ethnic groups in the country. They didn't like the Chinese (particularly the merchants), they considered the Cambodians to be less advanced, and they showed their contempt for the Montagnards by calling them *moi* [savages or barbarians]. They did not, however, bad-mouth the Nungs—that would have been hazardous to their health!

Major Tu had not been happy with the overall performance of the 5th ARVN Division, and he didn't think that it was killing enough Viet Cong. He went to the commanding general and convinced him to allow Tu to turn the Recon Company into a commando-type unit. (When you're buddies with the president, it doesn't take much to get your boss to go along with your ideas.)

The first thing he did was build up the company to about twice the normal size, and he sent it through the ARVN Ranger course at one of the training centers. Then he began to operate it as a separate battalion, but with one major difference—it seldom had to hump the bush looking for the Viet Cong. The Recon was sent into known VC areas, where the probability of contact was almost 100 percent. If one of the regular infantry battalions ran into something and the Recon was available, the battalion was ordered to become a blocking force, and the Recon took over the attack. The whole thing was immensely successful, and the company's kill ratio was about fifteen to one—that is, it killed about fifteen VC for every man that it lost. No doubt the ratio was much higher because the Viet Cong were able to carry off some of their dead and because some of their wounded died later. Body counts had always been suspect, but these figures were the ones which could be confirmed by the company advisors. (I have no idea what MACV headquarters did to the numbers when they finally arrived in Saigon.)

These weren't mindless suicide missions. As the magazine article had said, Recon soldiers were armed to the teeth. Every weapon in the company was fully automatic. Unlike the standard ARVN unit, these guys had M-16s and fully automatic M-14 rifles, and each platoon had at least one M-60 machine gun. They carried twice as many M-79 grenade launchers as a U.S. rifle company did, but unlike the Americans, the grenadiers also usually carried M-16s. On top of everything else, most Recon soldiers carried M-72 LAWs (Light Antitank Weapons) and knew how to use them. Besides its own fire power, the company could usually count on priority artillery, gunship, and air support. Considering everything, they were more than a match for any comparably sized VC or NVA unit.

However, the success that came from turning the Recon Company into a small assault battalion did create a problem—it couldn't fight and still gather intelligence. Instead of further increasing the size of the company, Major Tu decided to create a completely separate reconnaissance unit—the G-2 Recondo Company.

Although there were many similarities between the Recon and the Recondos, such as aggressiveness and weaponry, there were some very big differences. Most of the G-2 Recondos were "cowboys" and criminals. The cowboys were juvenile delinquents who had been caught and forced to join the army. The criminals were just that—men who had been convicted of crimes that ranged from armed robbery to murder. Major Tu had recruited them directly from prison, just like in the movie *The Dirty Dozen*. If they would fight, he offered them freedom and an opportunity to continue a little larceny on the side. *

They didn't much give a damn about Vietnam, but they really got off on being in a company like the G-2 Recondos, and they considered themselves to be comparable to the *Légion Étrangère*—the French Foreign Legion. Since they had such a bloody reputation with the Viet Cong, they figured they'd be done for if North Vietnam won the war. Almost to ensure that they wouldn't survive, they all wore a small triangular metal badge with a skull and crossbones and the words, *"Sat Cong"*—Kill Communists. Some of them had the motto tattooed on their arms or chests.

The Recondos might not be soldiers serving their country, but they had absolute loyalty to their leaders, their advisors, and to their sister company, 5th Recon. The Recondos were organized into three platoons of about thirty men each. One platoon would send out three long-range patrols, usually dressed in VC uniforms and carrying AK-47s, to gather intelligence and play mind games with the VC. For example, a patrol might link up with a band of Viet Cong, pretending that it was another VC unit. When the VC were off guard, the Recondos would kill them all and leave little printed cards showing

---

* In one after-action report, I described how the Recondos had aggressively closed with the Viet Cong and wiped them out in hand-to-hand combat. At the end of the report, I made the statement, "When the G-2 Recondos had killed all the VC, they began looting the *friendly* village in earnest." The report was returned to me, and I was asked to remove that particular sentence—I refused.

they had done it. Occasionally they would leave one Charlie wounded so he could tell his buddies that his unit had been knocked off by ARVN dressed as VC. As the story got around, it created fear and suspicion among the Viet Cong.

The two platoons that were not in the bush were available to assist a patrol in trouble or to respond quickly to new reconnaissance or reaction missions. They were also available to reinforce the 5th Recon Company when it had a good fight going.

Gene and I had been talking for a couple of hours when the rest of the team walked in the door. The first guy Gene introduced was a tall, lanky lieutenant with a crooked, boyish smile. Lieutenant Tony Callan was the assistant senior advisor, and he had primary responsibility for the Recondos. He wore infantry crossed rifles, but he was actually a Military Intelligence (MI) officer. Under normal circumstances I would have had little use for an MI officer, but Tony turned out to be more infantry than many infantrymen I knew. He had asked for MI in college without really knowing anything about it. Now, after serving as a combat officer, he wanted to be infantry. Standing there with his hands in his pockets, Tony reminded me of a schoolboy being introduced to his new teacher.

The next man was a skinny dude wearing glasses. He had on a set of camouflaged tiger fatigues—and a damned camouflaged *cowboy hat*! He wasn't exactly ugly, but he certainly would never be selected to appear on an army recruiting poster. He was about thirty years old, so I knew that he was probably a sergeant first class even before Gene told me. His name was Glen Septer, and he was the team's senior NCO. Unlike Lieutenant Callan, Septer was being very outgoing and talkative, but he too was sizing me up.

The last guy was a black man about my height, who looked to be in his mid-twenties. Staff Sergeant "Pat" Patterson was the team's other NCO. He didn't have any unusual physical characteristics and just came across as a good, solid soldier.

Captain Bolin told us that the two companies were having his farewell party that evening and that we should watch our liquor because we were scheduled for an Eagle Flight the first thing the next morning. Since I had never run that kind of operation before, he would go along to show me how to do it. After that the team would be all mine.

Any time a new man joins a team, everyone wants to know as much as they can about him—particularly if he is going to be the boss. The first few days are always a little touchy. The men want to know what kind of man and leader you're going to be, and they want to impress you with how good they are. We all had a big advantage in this game, because everyone knew of the other's reputation. They had made certain to check me out and seemed satisfied that I knew my business. When Colonel Sonstelie told me about my new job, I did the same thing and found out that they all had outstanding reputations. The Vietnamese company commanders had talked to Ninh and his officers about me, so I wasn't an unknown entity to them either. The only thing that remained to be settled was how our personalities were going to mesh.

Bolin's party was a real bash. The Recon Company had its own compound just across the road from the division's main gate. Although it had berms, bunkers, and barbed wire, it wasn't anything like Paris Tan Quy or, thankfully, An Son. It had tin-roofed, wooden barracks with concrete floors. The troops had U.S. Army bunks and even a recreation room with a TV. I found that they were paid much more than the average ARVN soldier, and they received bonuses when they fought a particularly outstanding battle. In addition, the company kept most of the VC weapons it captured and used them for trading with the Americans to get things that the unit could use. Their comparative wealth was obvious from the party food and the many bottles of hard liquor on the tables.

The G-2 Recondos lived in what had been a World War II Japanese Army prisoner of war camp at the far end of the division compound. The troops slept in the old concrete POW barracks inside a high-walled compound. I found their accommodations a little unsettling, but the Recondos were happy, probably because it was much better than the Saigon prison. (The Recondos were not prisoners; the compound was just conveniently located.)

Gene introduced me to Captain Hung, the Recon Company commander, and the other officers and senior NCOs in both companies.

I shook hands with everyone but stayed in the background, because it was Bolin's party and because it was not the time for me to make my presence felt. It was just as well, because the party gave the soldiers an opportunity to check me out personally—and they did. All evening I was surrounded by men who wanted to drink toasts with me. Actually, they weren't toasts—they were drinking contests. A soldier would come up, hold out his whiskey, and challenge me to beat him to the bottom of the glass. Other soldiers would stand around us, chanting "Yo, Yo, Yo," as we gulped the liquor down. I wasn't much of a drinker, and so everything became blurred pretty early in the evening.

Gene shook me awake early the next morning. God, did I have a hangover! I took a handful of aspirin and slowly dressed, trying to move my head as little as possible. We grabbed a quick cup of coffee and headed over to the airfield to meet the Recon.

An Eagle Flight is basically an aerial search and destroy operation. You loaded about eighty troops into ten or twelve UH-1D slicks, but you left them and two helicopter gunships sitting on the airstrip. Then you took the other two gunships and a C&C bird (a command and control Huey) and flew around looking for somebody to kill. If you spotted some VC or a suspicious area, you used the gunships with you to try to pin them in position while you radioed for the slicks to bring in the troops. The two gunships that had been working over the area were replaced with the two that were escorting the slicks, and they went back to rearm and refuel. You, the Recon Company commander, and the guy who owned the helicopters flew around over the LZ and orchestrated the whole thing. If the fight got good, you could bring in more troops with your slicks.

If a suspicious area turned out to be cold, you brought the Hueys back and picked up your ground troops. Then you went off hunting Charlie somewhere else. Eagle Flights offered a lot of flexibility, but you had to keep track of all kinds of technical details, like fuel levels, engine temperatures, ammo status, flight times, artillery, and a hundred other things—but when you had it all under control, it was beautiful, and it gave you a hell of a sense of power. The Vietnamese company commander ran the ground troops, and the advisor ran just about everything else.

Since Gene Bolin was in charge, I just sat back and watched what he did. Having nothing to keep me busy, I began to feel the chopper's vibrations mixing up the stuff in my stomach. No matter how much I tried to concentrate on the Eagle Flight, the nausea kept building until finally I couldn't hold it any more. I yanked off my soft Ranger patrol cap and puked my guts into it. Captain Hung and Gene looked around and started laughing as I sat there holding a hatful of vomit. I was embarrassed, and the smell was filling the cabin, even with the two side doors open. Finally I leaned over and threw the thing out the door. I had expected it just to fall away in the main rotor's down-wash, but I hadn't considered the wind created by flying at eighty knots. The hat only got about six inches beyond the door when a horizontal blast of air hit it and sprayed about a quart of puke all over the door gunner. He had his helmet visor down, but I could see the horror and hate in his eyes through the streaks of vomit. It isn't smart to screw with a guy holding an M-60 machine gun, so I climbed over Hung and Bolin and took a seat on the opposite side of the aircraft. They hadn't seen what happened, but they heard the door gunner screaming at me, so my little accident wasn't secret for very long. When we landed back at Lam Son, I high-tailed it through the berm as fast as I could move. I would have apologized, but the door gunner had a .45-caliber pistol in a shoulder holster, and I wasn't going to take a chance that he might use it.

I spent the next few days getting to know the three guys on my team and the Recon and Recondos. I was really impressed with the two companies, but I found that I had inherited a badly shot-up outfit. Only a week before I joined the group, the Recon Company had lost forty-two men killed or wounded at a place called Bo Duc, which was near the Cambodian border. It had been sent up there to reinforce a district headquarters against an expected Viet Cong attack. It only had a couple of days to work on the compound's defenses before it —and a 100-man RF company—was hit by the entire 272d Viet Cong Regiment, a force of about two thousand VC. It had been a hell of a fight—the VC twice overran half the compound and the Recon twice drove them out in hand-to-hand combat. When dawn broke seven hours after the initial attack, the Viet Cong withdrew, leaving ninety-six bodies and sixty-two weapons behind in the compound and hanging on the wire. Several battalions from the 5th ARVN and 1st U.S. Divisions pursued the 272d VC to the Cambodian border, and there they found mass graves with about 300 more bodies of Viet Cong killed at Bo Duc. The "bean counters" declared it a big victory—seventy-five Recon and RF casualties compared to about 400 known VC dead and untold others killed and wounded.*

While statistically it was a victory, we had still lost a lot of almost irreplaceable people. Because the Recon was now so understrength, we were only able to operate for the rest of December by using the available Recondo platoons to augment the company. Even then we could only field between 80 and 100 men. The division was giving basic training to 115 new Nung recruits, but until we got them and gave them additional training, we were pretty limited in what we could do. It was just as well though, because the company needed some time to lick its wounds before going back to fight large Viet Cong units.

On 29 December, after having conducted several more small one- and two-day operations, we were told that the 4th Battalion, 7th Regiment had hit a large VC force and had taken a lot of casualties. General Thuan ordered us to gather all our available troops and be prepared to go in and take over the fight at first light the following day. After alerting our people to get ready, Hung and I headed over to the division operations center to get briefed on what had happened and to find out what assets we were going to have.

We were going to be the division's main operation that day, so it allocated us all the choppers and airstrikes that it had been given by III Corps headquarters: one lift-helicopter platoon, two light fire teams (UH-1C gunships), and two air strikes. We had one battery of 105mm artillery in direct support and an additional battery on call, if needed. We were told that we could expect a couple of more airstrikes if we really got into something. I was satisfied that we had enough to start with, and I hoped there would be more if we needed it.

---

*The company was recommended for the U.S. Presidential Unit Citation, the highest combat award a unit can receive. Probably because they were Vietnamese, the award was downgraded to the second highest—the Valorous Unit Citation.

The 4th Battalion had been running a search-and-destroy operation near Ap Nha Viec. It had crossed Road 8-A and was moving across a 300-meter-wide field when at least a company of VC opened up on it from a tree line. Other than dry, knee-high grass, there had been no cover in the field, and the battalion had lost a lot of people before it was able to withdraw to the other side of the road. Though it was now dark, the VC were still periodically shooting at the battalion's troops to keep them from recovering their casualties or taking any offensive action.

After being briefed, Hung and I sat down to plan how we were going to conduct the operation. We searched the map for a good place to insert the company. We hoped that we could hit the Viet Cong from the side or rear, but except for the area south of 8-A, there was no place we could put in a complete lift of eight slicks. This meant that we would have to land in full view of the VC, who were about 600 meters away. We'd need to put some fire on Charlie while we air-assaulted in. We worked out our timetable and gave a copy to the operations people so they could schedule our support.

At dawn the next morning, the Recon and Recondos were formed and ready to move. I couldn't help noticing the contrast between the two units. The Recon soldiers were uniformly dressed and equipped and standing in paradelike formation, looking every bit like the professionals they were. The Recondos were standing in what could only be described as a "group" and were wearing a variety of uniforms, mostly tiger-striped, camouflaged fatigues. Many of them wore cloth cowboy hats. Where the Recon soldiers carried M-16s and M-14s, the Recondos were armed with a variety of weapons, from Thompson submachine guns to AK-47s. The only thing that they all had in common was a small black and red neck scarf that they wore for identification. (The Recon wore yellow neckerchiefs for the same purpose, and we often compared ourselves to the horse cavalry riding to rescue the settlers.)

At first the operation went according to our plan. The artillery pounded the tree line until the fighter-bombers arrived. When the pilots were ready, the U.S. Air Force FAC had them put "snakes" and "nape" (bombs and napalm) on the VC positions. When they were done, the gunships covered our landing with half the force. We un-assed the choppers, ran to the road, took up positions, and started firing into the tree line. A short time later the gunships expended the last of their ammunition and left to rearm and refuel, and the artillery started back on the target.

Twenty-five minutes later the second lift, with the rest of our troops, started into the LZ, while we covered the landing along with the gunships. They joined us at the road, and we got ready to assault across the 300 meters of open ground. Just as Hung was about to give the signal to move, the FAC called on the radio and said that the second flight of two F-4 Phantoms was on station and ready to put its bombs in. Hung and I looked at each other with surprise—this airstrike was supposed to be on call. Now that the F-4s were here, we either had to send them home or delay our attack until they were done dropping their ordnance.

I told the FAC that he hadn't read our time schedule very well, because we were ready to launch our attack. He didn't like my comment and heatedly told me that if we sent the planes back, we wouldn't get any more, even if we needed them. Hung shrugged his shoulders, and I told the FAC to "put 'em in."

For some reason the first pass was too far beyond the tree line, and the bombs exploded harmlessly away from the VC positions. I yelled in the radio handset and told the FAC that he had missed the target completely. As though he was lecturing a small boy, the FAC responded that it was a good hit because he was certain that Charlie had moved. I hit the roof! "Listen you fuckhead! *I'm* the ground commander and *I* decide where the shit goes! Put the next pass on the tree line, where I told you to."

A highly pissed FAC then told me that I couldn't talk to him that way—he was an "Oscar Five," meaning he was a lieutenant colonel.

"Hey, asshole. I'll talk to you any way I want 'cause you work for me, and that gives me all the damn rank I need."

He didn't respond, but the next pass was on the tree line. While the airstrike was going in, the light-fire team called and said that the gunships were starting to get low on fuel and could only remain a few more minutes on station. I radioed the FAC and asked him how much longer his F-4s were going to need to expend the rest of their bombs. Still angry, he snapped back that it takes a while to use up

150

the bombs and the 20mm cannon ammunition. Calling back, I told him, "Negative on the twenty mike-mike [20mm]." I didn't want to waste time while they strafed the tree line. I had two gunships with 2.75-inch rockets and machine guns that I had to use while I had them.

This guy was probably some old wide-assed desk jockey who had decided to leave the paperwork in his in-box for the day while he went out to play forward air controller. He gave me an argument that he had to use all the ordnance—or else. Or else what? I had taken all the bullshit from this guy I was going to.

"Listen Mr. Oscar Five, you pack up your little air show and hit the fucking trail. I don't need you to tell me how to fight the war, and I don't ever want you coming out to support me again. If you ever show up over me again, I'll have my people shoot your ass out of the sky."

Without responding, he kicked his Bird Dog over into a hard turn and took off. * Just as Hung was signaling the troops to move forward, the gunships called and said that they were "bingo" on fuel and had to leave us. Great! The damned FAC had not only wasted bombs, but also the only air cover we had on hand to help us cross the field. When I told Hung, he didn't show much anger, just resignation.

We moved as quickly as we could, firing from the hip and trying to get across the 300 meters of open ground before Charlie could react. As I was running, I kept passing dead ARVN soldiers who had been killed the previous day—they were everywhere. The 4th Battalion had really taken it in the shorts.

It took the VC a couple of minutes to react, and by the time they began firing, we were only about fifty meters from their positions. We roared into the tree line, and in a few minutes we had killed the fifteen or twenty VC there. The others had either pulled out during the night or had been killed by the artillery or the aircraft. We only had a few men wounded, and by late afternoon we were calling for the slicks to pick us up.

---

* I used a lot of airstrikes in my two years in Vietnam, but this was the only time that I ever had trouble with a FAC. All the others were great, and we loved both them and their fighter jocks.

When we got back to Lam Son, I shook hands with Hung, gave a thumbs-up to the troops, and walked back to Gosney Compound with Glen, Tony, and Pat. After dropping my web gear and rifle on my bunk, I went to the officers' club. I had just finished a beer when Colonel Sonstelie walked in. Spotting me, he made his way over to the bar. After saying that he'd heard that we'd had a good operation, he said, "I had an interesting visit from an air force lieutenant colonel this afternoon. It seems he wasn't very happy with some of the comments a ground commander made to him."

When I explained what an abortion the airstrike had been and how we had had to cross that field with no gunships because of his incompetency, Sonstelie grinned and told me not to worry about it. He said that he'd reminded the colonel that the ground commander was the man in charge—and then had thrown the man out of the office.

12 NEW YEAR

I'd only known Captain Hung for a few weeks, but I liked him a lot. He was a tough commander, but he was still able to laugh. He was a little taller than the average Vietnamese and more solidly built. He was a handsome man, and there was something about his features that suggested there might be some French blood in his ancestry. He had dark, penetrating eyes and a small round scar on each side of his neck, where an AK-47 bullet had gone through his throat just below his Adam's apple. One day when he took off his shirt, I saw several other scars where he had been wounded. One, which was about six inches long, had obviously been serious. This guy had been in some heavy combat. I was a little embarrassed, because despite almost eight months in the field and a lot of firefights, I didn't have a scratch—except for those caused by a couple of punji stakes, scars that I wasn't particularly proud of.

Things were unusually quiet in January, and very few contacts were being reported anywhere in the division's AO. It was just as well, because we had gotten our new Nungs and were in the process of giving them advanced tactical training. Although they had only had

153

about six weeks of basic training, they looked sharp, tough, and ready to go. If we could keep out of any big battles for a few weeks, we'd have time to bring them up to speed on the way we worked.

One of the things we had to do was to train them in how to use our weapons. All the rest of the division had World War II stuff, but we were armed with the new American weapons: M-14s, M-16s, and M-60s. About a third of our people had the big, heavy M-14 rifle, and we advisors were working to replace them with the smaller, lighter M-16s. Since the U.S. didn't have any immediate plans to give the new weapons to ARVN infantry units, Bolin and the team had worked out a system to get them—*they stole them.*

Bolin had told me how it was done, but I decided that I would like to see it for myself. Glen tried to discourage me, but I insisted, so one morning he showed up in front of my hootch with a jeep and a canvas-covered trailer. Glen lifted the cover and showed me a pile of VC weapons: AK-47s, SKS carbines, and some old bolt-action French rifles. Reaching inside his shirt, he pulled out a couple of big VC flags that were slightly dirty and bloodstained. He had them made in a local tailor shop, and he used chicken blood to give them a look of authenticity. We climbed into the jeep, and still complaining that this wasn't "officer business," Glen drove out of the compound.

Glen and Pat had the process down to a science. First we drove to the property disposal yard, and after a private conversation with the sergeant in charge, we unloaded the VC weapons and gave him the VC flags. While two of his assistants were stashing the VC loot, the man guided our jeep to an area next to the fence, then pointed at a pile of U.S. weapons with turn-in tags tied to their barrels. Glen told me only to take the ones whose tags had a red check mark on the back. They were the ones that could be repaired. After we loaded all the M-16s and M-60s that the trailer could hold, we drove to a maintenance outfit. After Glen explained to a suspicious sergeant that I was "one of them," we unloaded the weapons that we had picked up at property disposal. Then we loaded a bunch of the rifles and machine guns which the maintenance people had repaired. When we were finished, we tied the trailer cover down to conceal the cargo and drove back to Lam Son.

Glen explained that he and Pat went through this procedure every couple of weeks. Each time they were able to get twenty or twenty-five M-16s and usually one or two M-60 machine guns. In another six or eight weeks we should have an M-16 for every one of our new troops, so that we could get rid of the automatic M-14s. I didn't feel uncomfortable with our illegal activities because we weren't personally benefitting. We were trying to arm our soldiers with weapons that gave them as much firepower as the enemy. Later I made certain to periodically report the numbers and types of weapons the companies had so no one could accuse me of running some sort of private black market operation—which of course I was.

Shortly after New Year's Day, Pat came to me and asked to be transferred to a REMF job. He had been humping the bush for a long time, and the big battle at Bo Duc had convinced him that he had better get out of the Recon if he really wanted to go home in one piece. He assured me that his request had nothing to do with my taking over the team, but he was tired and getting too shaky on operations. I understood and asked Colonel Sonstelie to give him a safe job for his last month in Vietnam. We were all volunteers, and we could "un-volunteer" any time without any questions. There was no disgrace in Pat's asking for a different job.

In a few days Division offered me several NCOs, and I picked the one I thought would best fit into the team. Staff Sergeant Chuck Zollenger was a good-looking guy in his mid-twenties, who had just the right personality and experience.

Like the rest of us, Chuck was a "laugher," and we all hit it off right away. Glen began training him for his new job by taking him around to meet the various people we traded with, the people who supplied us with the materials the companies needed. When the two of them got ready to go to Cu Chi, I decided to go with them to try to track down some of our wounded. They had been evacuated there, and we didn't want them to get lost in the U.S. casualty system.

We had two choices for transportation to Cu Chi—hitch a ride in a chopper or drive over on 8-A. Since we couldn't scrounge much when we were bumming around in a helicopter, we decided to take the jeep. Once there had been a bridge across the Saigon River at Phu Cuong, but the VC or Viet Minh had blown it up. Now we had to cross the river on a military ferry—a floating raft powered by a

155

couple of big outboard motors. There was some civilian traffic but not much military traffic on 8-A, so the VC only mined it occasionally. About a kilometer west of the river, the trees and brush gave way to a wide-open expanse of now-dry rice paddies. There were some spots of heavy brush about four or five hundred meters north and south of the road, but there were no good places for the VC to hide an ambush force close to the road.

Glen drove as fast as he could for the first few hundred meters beyond the river bank, and then he put the clutch down and let the jeep coast up to the edge of the trees. As soon as we broke into the open, he shifted into low and gunned it. There always seemed to be at least one Charlie in the tree line to the south, somebody who delighted in taking potshots at military vehicles driving down the road. The latest sniper was a lousy shot, particularly at that long range, so no one bothered much with him. The general feeling was that if we killed him, the VC might replace him with someone who could shoot accurately. By coasting up to the edge of the trees, we made sure he wouldn't hear us coming. We could catch him off-guard and be halfway to the trees on the far side before he could take a shot at us. By that time we'd be doing thirty-five or forty miles an hour.

At first none of us could tell if he was shooting at us because of the engine noise, the wind, and the shock of bouncing over the potholes in the road. However, when we had to slow down for a particularly large hole, we saw a round kick up dirt right behind us. The bastard was there, but he hadn't gotten any more accurate since the last time I had made this run. Glen shifted into a lower gear, and we literally flew over the dip in the road, then quickly shifting into high, he accelerated back up to forty miles per hour. In another thirty seconds we passed a PF (Popular Force) outpost and were out of range. The rest of the trip to Cu Chi was quite a bit more sedate.

We stopped by the hospital, and after a little hassle, we tracked down five of our six wounded soldiers. They were recovering pretty well and were in "hog heaven," laying in a nice clean hospital ward. They'd even developed a taste for American food, and I thanked the medics for taking care of them.

One of the troops had died, so we had to drive over to graves registration to locate his body. To reduce the effect on morale, the

Americans had built a high wooden fence around the place, but everyone knew what went on inside. I hated going there, but at least the people in division graves registration units were appropriately solemn. I had been to one in Saigon and came close to blowing the damned ghouls away. They were laughing and scratching and playing rock music as loud as they could on their portable radios. Maybe they were like that to keep death at a distance, but the guys in the body bags deserved a lot more respect than they were getting.

A serious guy, acting like an undertaker back in the States, had to unzip four bags before I was able to find our man. He'd been pretty badly fucked up, but I identified him by the "Sat Cong" tattoo on his chest. I told them that I would have someone from 5th ARVN come over in a day or two to claim the body.

When we came out of the building, I noticed two big piles of uniforms and boots. Much of the clothing was bloody and shredded, but most of the jungle boots looked to be in pretty good shape. I thought about scrounging some of the smaller sizes for my Vietnamese troops, but I decided that I'd just as soon not have them wearing dead men's boots. We'd get them new ones, or they could continue to wear the ones they'd been issued.

After collecting a trailerload of miscellaneous supplies, we headed back to Lam Son. This time I don't think Charlie shot at us because we hit the open area during siesta—he was probably snoring away in his hammock.

Early in January Major Clark, the G-2 intelligence advisor and my new boss, and Colonel Sonstelie gave me a mission. General Thuan had decided that having a Ranger company wasn't enough; he wanted it to be airborne as well. At first I was excited when I heard about Thuan's idea. I didn't know what we would accomplish by being able to parachute down on the VC, since it wasn't a very effective way to insert troops, but I liked the added "elitism" and the additional jump pay. Before I had time to relish the idea, Clark and Sonstelie hit me with the punch line—my advisor team would give the training. We were to train not only the two companies—but also *General Thuan* and some of his top officers!

If anyone says that training a soldier to parachute out of an airplane involves a lot of technical stuff, he's lying. There really isn't much to it. You explain a little about how the parachute works, how to put it on and adjust it, how to hook up and then jump out the airplane door, and how to *try* to control the parachute during the descent. Finally you teach him how to hit the ground, hopefully without busting his ass. The U.S. Army's course is three weeks long: two weeks involve ground training, and during the last week you make five jumps. During ground school, teaching the various techniques only takes a few days. The rest of the time is spent repeatedly going over them to make sure everything is instinctive. There isn't a lot of time to think from the point you leave the aircraft until you hit the ground. Of course, the program includes lots of unnecessary harassment and physical training, too.

I wasn't worried about giving the training because everyone on my team was airborne qualified, and both Glen and I had served quite a few years in airborne units. But I knew that if General Thuan took the course, every damned general and colonel in MACV would be checking on me to make certain I didn't kill him. I argued that we were fighters and not trainers, but I got nowhere. I was told that the division's engineers would build any facilities I wanted, such as swing-landing trainers and PLF (parachute landing fall) platforms. Division would get the parachutes for us, arrange for the jump aircraft, and provide security around a drop zone, as well as any other support that I needed. Handing me a copy of the manual, *Technical Training of the Military Parachutist*, Colonel Sonstelie said, "I know you'll do an outstanding job."

I went back to Gosney Compound and told my team about our new mission. Just as I had, the men grumbled and complained. Their pissing and moaning did die away some when I reminded them that we'd all start drawing jump pay when we made our first jump. When I told them to go buy red airborne berets and ARVN jump wings, they finally came around.

I worked up a training program and some simple lesson plans I knew the visitors would undoubtedly ask to see. It wasn't the U.S. Army Infantry School, but it was enough for Vietnam. I gave Major Clark some sketches of the things I wanted the engineers to build, and I sent the team off to scrounge some of the equipment we would need, like parachutes and harnesses for ground training.

While we were stuck in base getting the training organized, Captain Clarence Hooker invited me to ride along on one of his "last-light" flights. Hooker was the chief of the small aviation detachment at Lam Son, and his team of four officers flew O-1 Bird Dogs, supporting division operations. They flew reconnaissance, adjusted artillery, directed helicopter gunships, and generally helped out however they could. They were worth their weight in gold to us guys on the ground, and they repeatedly took their little airplanes through all kinds of shit, armed only with a few 2.75-inch white phosphorus (WP) rockets and an M-16 wedged next to the pilot's seat. They were forbidden to carry high explosive rockets, because the army didn't want them to start acting like fighter jocks in an airplane that only flew about ninety knots.

Hooker was a little crazier than the other pilots and took a lot of unnecessary risks, but he enjoyed himself. When he asked me if I'd like to go up with him, it only took me about ten seconds to grab my rifle and a bandoleer of ammunition. On the way out to the plane, he showed me on the map where we were going to recon: Ap Nha Viec, the Hook, and the Iron Triangle—some of my favorite places.

Every evening, an hour or two before dark, Hooker and his people would fly over parts of the division's AO—thus the name "last-light" flight. They were looking for any VC activity that might indicate what Charlie was going to do that night. They seldom saw anything more than a couple of VC, but they could often pick up signs if something unusual was happening. Anything out of the ordinary was checked and reported to the division TOC. Sometimes they spotted one or two VC in a free-fire zone and blasted them with a WP rocket or called in artillery.

Compared to busting brush on the ground, I thought that this flying business was luxury. You didn't have to hump a heavy rucksack or get your boots dirty, and except for an occasional .51-caliber heavy machine gun, you didn't get ambushed. On the other hand, if the VC hit you, it was a long fall and a sudden stop when the plane hit the trees.

Since I was familiar with the ground we were covering, I had Hooker check out a couple of places I knew the VC regularly used. The first few locations were dry and we couldn't see any activity, even though we flew just above the trees. Shortly after we crossed the Saigon River into the Triangle, Hooker noticed some hazy smoke in a small clearing and began circling over it. It took a few minutes, but we finally spotted a little camouflaged hootch where some VC were apparently cooking their evening rice. I stuck my M-16 out the window and emptied a magazine into the area. This was the first time I had ever shot at anything from a moving aircraft, and my rounds went everywhere, except where I was aiming. I'd aim at the hootch and see the bullets kicking up the dirt more than fifty meters away. Hooker reminded me over the intercom that I had to adjust my aiming point, because the plane was moving along at eighty to ninety knots. It took me several magazines to get the hang of it, but I was finally able to put most of my rounds on the target. I didn't think that I was doing much damage, so I asked Hooker to fly a tight circle directly over the target; that way I could drop a couple of hand grenades on the place. Hooker wasn't certain that he liked that idea, so he leveled the plane and climbed to a higher altitude. When we were high enough to be safe from everything but a .51-cal, he went over the procedure I was to use. He wanted me to stick the grenade out the window *and then to pull the pin*. He didn't want to risk my dropping a live grenade inside the aircraft and having only four-and-a-half seconds to find it and chuck it out the window. I could agree with that logic, so I got ready and Hooker took the plane back over the hootch. Just as they must have done it in World War I, I held the grenade out window and released it when he said "Drop!"

After I used up my four grenades, I stuck my head out the window to try to see what damage we had done—it didn't look like much. *Oh well, it had been fun anyway.* When we got back to Lam Son, I told Hooker that I'd get some stuff to take with us the next time. I'd load a bunch of magazines with tracers so I could see where my rounds were going, and I'd bring along a bag of grenades. I'd also scrounge some 2.75-inch HE (high explosive) rockets so we could replace the normal white phosphorus marking rounds with something that could do some damage. Hooker's people wouldn't tell on him, and we'd shoot off the unauthorized rockets before we landed back at the airstrip.

A week later I went on R&R in Hawaii. I hadn't seen my wife in eight months, and since the Recon wasn't doing much besides training the new Nungs and getting ready for airborne training, it was a good time to go. I did all the normal things a married man does when he meets his wife in Honolulu—plus something she thought was a little unusual. I spent one afternoon buying stuff that I wanted to take back to Vietnam. I went to a gun shop and bought a rifle scope and mount for my M-16 and a couple of clips for the .32-caliber pistol I always carried in my pocket. Then I went to a hardware store and bought a hand drill, a small wrench, screwdrivers, and some nuts and bolts to mount the scope. Finally I spent several hours wandering around a sporting goods store, buying whatever I thought I could use in the bush. Civilians certainly had better equipment than soldiers.

When I got back to Lam Son, I couldn't wait to get to work, and I had barely taken off my khaki shirt before I was drilling holes and getting the scope mounted on my rifle. The staff officers in my hootch thought that I was nuts, but I was happy and became ecstatic when I got the thing zeroed-in the next morning. Now I could do some real work, rather than just spray the countryside with automatic fire.

The team had gotten almost everything ready to go while I was in Hawaii, to the point that we could start the training in a few days. However, Tet was only a week or so away, and many of my Vietnamese troops would be going on leave to visit their families. I went to Major Clark and Colonel Sonstelie and asked that we delay the start of the course until after the troops came back. I didn't want to see them get halfway through ground school—only to have them forget what they learned while on leave, where they would be getting drunk and having a good time. After checking with General Thuan, the delay was approved—we would start right after the Tet truce.

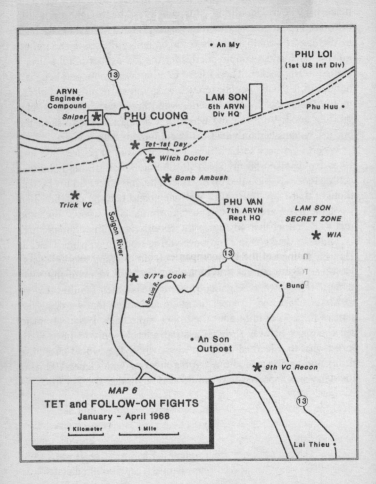

MAP 6
TET and FOLLOW-ON FIGHTS
January – April 1968

1 Kilometer   1 Mile

An My

PHU LOI
(1st US Inf Div)

13

ARVN
Engineer
Compound

Sniper ✱

PHU CUONG

LAM SON
5th ARVN
Div HQ

Phu Huu •

✱ Tet-1st Day

✱ Witch Doctor

✱ Bomb Ambush

✱
Trick VC

Saigon River

PHU VAN
7th ARVN
Regt HQ

LAM SON
SECRET ZONE

✱ WIA

13

✱ 3/7's Cook

Ba Lua R.

Bung

• An Son
Outpost

✱ 9th VC Recon

13

Lai Thieu •

13 TET—1968

What remained of the two companies (about eighty troops after Tet leaves were granted) was searching along Highway 13 a few kilometers north of Phu Cuong. The Tet truce was about to begin, and we had been ordered to make one last check of the area to make certain that the VC weren't getting ready to do something just before the cease-fire. It had been an uneventful, hot, dusty walk—the kind we hated. Normally, Highway 13 was always busy from 0900 in the morning until about 1700 in the afternoon, but today it was teeming with mostly civilians in Lambrettas, motorcycles, bicycles, and all sorts of other vehicles.

Tet was the most important holiday in Vietnam, sort of like Christmas, New Year's, and Thanksgiving, all rolled into one. It was a time for celebration and was always accompanied by noisy crowds, music, and unless the government prohibited them, fireworks. It was a time for the Vietnamese to be with their families, and custom dictated that they give presents wrapped in red paper to relatives and friends. It was obvious that most of the people who we were seeing were headed home for the holidays, because they had their red-wrapped gifts.

Everything seemed normal for the holiday season—except something kept nagging at me. Many of the travellers were young men of military age. This wasn't particularly unusual, because we had been authorized to allow fifty percent of our troops to go on leave, and throughout the Vietnamese Army, many soldiers were on their way to visit their families. What was unusual was that quite a few of the young men were in civilian clothes—not the normal dress for off-duty ARVN soldiers—and they weren't carrying Tet gifts. We joked about it and decided that many of the men that we were seeing were VC who had been given leave to visit their families. Had it not been for the impending truce, we would have detained them and turned them over to the police for checking. But we had been instructed not to take any offensive action, so we just noted the bastards and continued our walk in the sunshine. Although it bothered us, there was nothing we could do. Fortunately, in midafternoon we were ordered back to Lam Son. There would be no more combat until after Tet.

After breakfast the next morning, I laid down on the bunk and started reading a book. George was sitting next to me, picking at the hair on my leg and making little *ooh* sounds. Unlike caged monkeys, she kept herself pretty clean and got along well with everybody in the hootch—except the cleaning ladies. For some reason George didn't like females and was extremely jealous of any who came near me.

I was just getting to a good part in the book when the field phone rang. I picked up the receiver, and the guy on the other end identified himself as a TOC duty officer. He said that Colonel Sonstelie wanted to see me in the operations center as soon as possible. When I asked why, he only repeated that I was to come as quickly as I could. I thought that this was a little odd, but I picked up a new French MAT-49 submachine gun I had scrounged and strolled over to the center.

When I walked into the TOC, I found a bunch of Vietnamese and American officers huddled in front of wall maps and talking on radios. Colonel Sonstelie was standing next to the main operations map, talking to Major Clark. When he saw me, he motioned for me to join them. I walked over, but before I could say anything, he handed me a teletype message. It was marked "Top Secret" and addressed to every major commander in Vietnam. It was from General Westmoreland, and it said that intelligence reports indicated the Viet

Cong and NVA were preparing to launch large-scale attacks throughout the country in violation of the Tet cease-fire. Everyone was directed to make all preparations necessary to repel the attacks if they came.

I let out an "Oh shit!" and asked Sonstelie what he wanted me to do. He said that a similar message had been passed through ARVN channels and that Hung and the Recondo company commander should be getting their troops ready at that very moment. While they were doing that, the colonel wanted me to do a little job for him. He walked me over to another wall map showing the northern part of the division's AO and part of Cambodia. One of his captains pulled down an acetate overlay with grease pencil dots scattered all over the Cambodian side.

"This is a plotting of SLAR [Side Looking Airborne Radar] and Red Haze [infrared photography] reports from a couple of weeks ago. Each dot represents a VC or NVA cooking fire. The larger the dot, the bigger the unit. As you can see, the fires are pretty well spread throughout the area across the border. Now compare this plotting with the next ones, from more recent reports."

The next overlay clearly showed the fires closer together and nearer to the border. As he changed the acetate plots, you could almost see the VC and NVA units moving and concentrating just across the border. Then, the last couple of plots showed almost no cooking fires at all.

"As you can see, the VC and NVA moved right up to the border and then disappeared. It's pretty obvious that they have crossed into Vietnam, but we don't know where they went or what they intend to do. This, General Westmoreland's message, and our own intelligence, all point to something big about to happen."

Walking back to the big operations map, Sonstelie said, "Bob, I want you to get your team together and run a reconnaissance for us. Captain Hooker and his pilots are on their way to the airfield now, and I want you and your three people to fly over Binh Duong province to see if you can spot anything."

A few minutes later my team was standing at the map. I had divided up the area into quadrants, with the Saigon and Thi Tinh rivers forming the north-south line and 8-A the east-west one. I told them what I knew and assigned each of them a quadrant. Then we loaded into vehicles and drove the short distance to the airstrip, where Hooker and the other pilots were already warming up their Bird Dogs. Since Hooker and I were used to working together, I picked him to be my pilot.

He already knew what the mission was, so I only had to tell him which area we had. I plugged in my headset, and as we taxied to the end of the runway, we discussed what a waste of time this was probably going to be.

Once we got about 500 feet in the air, he pushed it over into a right turn and headed west toward Phu Cuong. In a few minutes we were over the main market, and Hooker turned south along the east bank of the Saigon River.

I was leaning out the right window when the plane suddenly banked, and I heard Hooker yell "Holy shit!" over the intercom. I turned my head and looked where he was pointing. Jesus Christ! It was some sort of Viet Cong parade! Hundreds of VC carrying AK-47s, RPGs, machine guns, and other weapons were marching down the middle of the street in a suburb of Phu Cuong—*in broad daylight*! About a hundred civilians were standing on either side of the street, watching the VC march by. Stretched across the street above the Viet Cong's heads was a white banner with red-lettered words. Although I couldn't make out what it said, it was obviously not government propaganda.

As Hooker opened up the throttle to get out of AK-47 range, I grabbed the push-to-talk switch on the radio and called Division. Breathlessly I told them what we had found and asked for an artillery fire mission. There was a long pause, and finally the division's radio operator responded, "Negative artillery."

At first I couldn't believe it, and then I got very angry. "Listen, you dumb shit! There are several hundred armed Victor Charlies marching right down a fucking street in the province capital!"

"I say again—negative artillery."

"Get the hell off the radio and give me someone who knows what he's doing!"

In a moment I heard Colonel Sonstelie call. "Tell me what you've got."

"I've got at least a full Victor Charlie battalion having some sort of parade. They've got weapons, and they're right in the southern part of Phu Cuong city at grid coordinates XT 809133. I need artillery *now!*"

"Now calm down. Are you sure they're Victor Charlies?"

I stared at the map in anger. The bastard had said, "Calm down"! I'd been in the field for *nine* months and in more firefights than I could remember, and he was telling me to calm down! Hooker had been monitoring the conversation and came on the intercom. "What's with that shithead? Maybe we can come in low and fast and snag that VC banner. Then we could take it back and hang it over his desk."

"Great idea, Hooker. How the hell are we going to keep from being blown out of the sky while we try to steal their sign?"

"Well, they haven't shot at us. Maybe . . ."

"No, Hooker. They aren't just going to watch quietly while we tried something like that. Besides, how would we snag it?"

"Yeah, I guess you're right. But why haven't they shot at us?"

"Beats the hell out of me. Maybe they're observing the cease-fire. Let's go back low and fast and see what happens."

Hooker banked the plane, and we roared back over the VC, but nothing happened. They were still marching down the street, and we could see that they were all looking at us, yet none of them made a move to shoot. Frustrated, we then did something really stupid—we made slow, tight turns about 200 feet over their heads. They just continued to look at us. Finally we came to our senses and realized that we were just dangling ourselves as bait—not a very smart move at all. I told Hooker to take us home.

Colonel Sonstelie kept calling me on the radio, but I refused to answer. Besides being pissed, I hoped he'd think that we might have been shot down. I wanted him to sweat. When we landed, I grabbed a jeep and roared off to the TOC. When I got there, I stormed in and immediately met Colonel Sonstelie. "I've been doing this shit for a long time, but since you apparently don't believe what I tell you, I think that it's time for me to get a new job."

Sonstelie put his arm around my shoulder and said, "I believe you, but we're in the middle of a cease-fire, and there's nothing we can do. I already reported it to III Corps, but they refuse to let us engage the VC unless they open fire first."

Sarcastically I replied, "Well, Hooker and I will just go back and get those bastards to shoot us down, and then you can do something."

"I understand how you feel, but I want you to get over to your unit and be ready when we need you."

"Colonel, I've seen one full battalion right in the city. If I've seen one, you can bet your ass there are a lot more that I *haven't* seen. When they hit us, we're going to be in deep shit. I've never fought in a city, but I expect that I'm going to take a lot of casualties in that house-to-house stuff. I'm not going to like knowing that I lost troops because nobody had the balls to do the right thing."

I drove over to the Recon Company compound. The eighty troops who had been held back from leave were already packed and ready to move whenever they got the order. More than half of them were the new Nungs, and this was going to be their first big fight. The damned VC had really caught us with our pants down with their Tet "cease-fire." Hung came over to the jeep, and I told him what I had seen from the Bird Dog. Although he was cool, he was also worried and angry, because he had been told the same thing—no action until the Viet Cong actually broke the truce.

I told him that I had to get the rest of my team and that I'd be back in a few minutes. As I drove back to the airstrip to pick up the others, it dawned on me that Hung's wife and kids lived in a little house not more than a couple of hundred meters from where I had spotted the VC battalion.

As my guys landed, I asked them what they had found. All said that there had been some signs of recent movement but that they hadn't seen any identifiable VC. When I told them what I had seen and that we had been ordered to sit tight, there was a barrage of angry arguments. I made no effort to justify or defend the orders, but I began to understand how Colonel Sonstelie must have felt. Pissed, we drove back to Gosney to pick up our gear and move it over to the Recon compound.

That night all hell broke loose. Shortly after midnight, artillery and gunships began pounding an area just beyond Lam Son's perimeter. At the time we knew that something big was happening, but we didn't get the full story until several days later. According to captured VC documents and POWs, the 2d Battalion of the Viet Cong's Dong Nai Regiment was preparing to attack 5th ARVN Division headquarters at Lam Son. As it was moving into its attack position, it passed through the area where the 273d VC Regiment was preparing for an attack on the large American base at Phu Loi. At 0100 a long-range patrol from the 1st U.S. Infantry Division discovered both units and called in artillery from Phu Loi. For the next thirty minutes, the Americans fired massed artillery on the VC. Then helicopter gunships began shooting up the retreating VC. The Dong Nai Regiment's 2d Battalion withdrew into the pro-government village of Phu Huu, about two kilometers east of Lam Son. The VC 273d Regiment pulled back into An My, another friendly village just three kilometers north of our compound. Apparently the Viet Cong thought that we wouldn't shoot at them while they were occupying pro-government villages—they were wrong.

At 0400, the Americans called in airstrikes and opened up with more artillery. At dawn 1st Infantry Division infantry and tank units swept through the villages and finished off the VC. According to official reports, they counted 440 VC bodies and captured nine wounded Charlies. The weapons haul was big and included three .51-caliber heavy machine guns, two recoilless rifles, and five 82mm mortars.

While these battles were just a short distance from Lam Son and we could watch them, they were just two of many in the area. All the radio frequencies were jammed with reports of heavy fighting. The VC had hit everything of any size—Binh Duong Province headquarters, the main police station, the ARVN engineer training center, and just about every major government installation in the southern half of the division AO.

There wasn't anything for me to do, except stand in the TOC and watch the reports being plotted on the operations map. There were grease pencil marks all over the thing. I put on my best "I told you so" look, but nobody paid any attention. They were too busy jumping through their asses to even notice that I was there. The situation was so chaotic that it would have been stupid to commit my Recon Company until daylight. Besides, General Thuan was not about to send out his only combat unit when the VC obviously intended to attack Lam Son. He needed us there to stiffen the cooks, clerks, mechanics, and miscellaneous "ash and trash" that were defending the perimeter. He alerted Hung to be ready to move at first-light, but until then, he was to have the Recon and Recondos standing by as a rapid reaction force.

As I was watching the situation develop, Hooker made his way through the crowd to where I was standing. "Jesus! The VC are blasting the shit out of Saigon. They're attacking all over the city, and the REMFs are catching hell."

"I'd like to see that. Those fuckers spend all their time partying, and now they can get a taste of what this war is all about."

Hooker thought for a moment and then said, "Why don't we take a Bird Dog and go for a ride over the city? It ought to be fun to see how bad it really is."

I didn't bother to answer; I just turned to Tony and told him that I'd be back in a couple of hours. "Hooker and I are going to fly down to Saigon to cheer for the VC."

With all the artillery and airstrikes around Lam Son, the only way to take off was to the south. We didn't know what might be there, so Hooker literally stood on the brakes while he revved the Bird Dog's engine to full power. When the RPMs were as high as he could get them, he released the brakes and we roared down the runway. He held the aircraft on the ground as long as he could to build up airspeed. Then he pulled back on the stick and we leaped into the air, climbing quickly to get above possible groundfire. When we leveled off at two thousand feet, we could see all of Saigon in the distance—explosions, tracers, flares, and fires everywhere. It was the biggest battle that I had ever seen, and it was a hell of a sight.

I heard Hooker over the intercom, telling me to switch to the other radio and to listen to the fight our guys were having at the golf course. I didn't know Saigon had a golf course, but two helicopter gunships were attacking a VC unit, and one of the pilots was reporting that he was taking heavy automatic weapons fire from the third hole.

The sky was so crowded with choppers and fighter-bombers that Hooker had to turn on his navigation lights to avoid a midair collision. Since we were merely spectators, Hooker kept our plane high, giving

us a panoramic view of the whole city. We were almost neutral—
we cheered our own airplanes, but we had to give Charlie his due.
He was doing a job on Saigon.

When our fuel got low, Hooker turned the plane north, and we
headed home. I had to get back to the troops and be ready to move
out when it got light. There were still a lot of explosions and firing
in Phu Cuong, An My, and Phu Huu, but Lam Son appeared quiet.
After we landed, I headed back to the TOC to find out what had
happened while we'd been gone. There was still a lot of confusion,
but most of the battles seemed to be under control—except for Phu
Cuong, where there was fighting throughout the city.

At 0700 I picked up the other guys, and we headed over to the
Recon compound. Hung was briefing his officers, and I listened as
he outlined his plans. There really wasn't much to plan. He was to
take the force into Phu Cuong, and he would get further orders as
the situation developed. When he was done, I said, "We don't have
a whole lot of soldiers to take on a city full of VC, do we?"

Hung gave me a weak smile but said nothing. I knew that he was
very concerned about his family but couldn't let that interfere with
his job. I hoped that we'd figure out a way to get into his neighborhood,
but I said nothing. There wasn't much that I could say, and I thought
it would be better to avoid the subject altogether.

At 0730 our well-armed, but under-strength, company moved
through the gate and turned down the road toward Phu Cuong. There
was a little joking to relieve the tension, but as we got closer, we
began to hear small arms fire mixed in with the heavier stuff and
everyone became grim-faced. It was pretty obvious that the VC were
there to stay and that this wasn't going to be a picnic. As we moved
into the eastern outskirts, it was like a ghost town. There hadn't been
any fighting here, but the people had all disappeared. We had been
spread out on either side of the road, but we pulled in closer as we
moved among the buildings.

As we carefully made our way along the main east-west street, we
began to see evidence of the recent fighting. There were few bodies,
but it was obvious that a lot of ammunition had been fired during
the night. A few hundred meters from the ARVN engineer compound
we began hearing tanks and some small arms and automatic weapons

fire. Most of the compound was protected by the usual high dirt berm, but on the side adjacent to the city, there was nothing more than a concrete wall that was about six inches thick and seven feet high. There weren't any mines or barbed wire on this side, so it shouldn't have been much of a problem for the VC to punch a hole in the wall and get into the compound.

When we got to the wall, the troops dropped into a shallow drainage ditch alongside the road and took a break. Hung and I spotted Colonel Quan and Major Stefaniw, Humphry's replacement. Quan reached out to shake hands, but I pointedly ignored him and walked over to Stefaniw. Quan was a little surprised at my obvious slight, but he recovered quickly and began to show Hung a map. I asked Major Stefaniw to bring me up to speed on what was happening. He explained that an unknown number of VC had taken the main two-story administration building and were holed up inside. The Vietnamese were trying to figure out how to get at them without destroying the building.

I suggested that we be given the job. The regular ARVN troops could give us covering fire while we made a dash across the hundred meters of open ground. Once we got to the building, we'd work our way from room to room, starting with the ground floor. If his men could get us some satchel charges, we'd make quick work of it. If not, we'd use hand grenades to clean them out.

He smiled, but said, "That's a good plan, but you'll blow the hell out of the building. The ARVN are trying to hold down the damage."

"This is the *engineer* compound. They've got all the stuff they'd need to repair any damage we'd do."

"Well, I don't think Colonel Quan will go along with it."

Just then, Hung came up and said, "Let's go back to the troops." He didn't appear very happy, so I asked him what the plan was.

"We just sit here until they figure out what to do with us."

When I told him what I had suggested to Major Stefaniw, he chuckled and said, "I made the same suggestion, and Colonel Quan said the same thing as the American major had."

We both laughed and flopped down in the ditch. It was their fight, so they could figure out how to get the VC without damaging their precious building. We would be content to watch.

172

I went over to a hole in the compound's wall to see what was happening. There were two tanks racing around in circles in front of the building, shooting at it with their machine guns. They kept moving fast in hopes that the VC wouldn't draw a bead on them with an RPG or a recoilless rifle. The only problem with that was they couldn't aim their machine guns very well and were spraying rounds wildly. As far as I could tell, the VC were perfectly safe. It was so stupid that I found myself starting to root for Charlie.

After a few minutes of watching, I thought I spotted some shadows in one of the second-story rooms. I raised my M-16 and turned the telescopic sight to full power. (I had gone first class when I bought the scope. It was adjustable from three to nine power.) Resting it on the concrete, I slowly scanned the upper windows. Every few moments I could see movement in some of the rooms. When I had one in the cross hairs, I squeezed the trigger. The shadow disappeared and I wondered if I had hit the guy. I watched the window for a few minutes and then went on scanning the others. During the next ten minutes I shot at five more shadows. All disappeared, but I couldn't swear that I had gotten any of them—at least I had a good time.

Hung yelled to me that we were moving out. We had been ordered to head toward the southern part the city, exactly what Hung had hoped for. We would have to go through the area where his wife and children lived. We saw no one as we carefully made our way down the streets. Everyone was either hiding in their house or had been able to get away from the area. We spread out on either side of the main north-south road, which was the section of Highway 13 that ran through the city. If this had been an American town, we would have been able to move along several parallel streets. But this was Vietnam, and it seemed that no two streets ran in the same direction for any distance. It was very difficult to maintain contact between the platoons because of all the small buildings and the alleys between them. It was also very tricky because we could be ambushed anywhere.

Every few minutes somebody began shooting. Either a Viet Cong would open up and our troops returned fire, or one of our soldiers would spot a Charlie and squeeze off a burst. Whenever the Recondos fired, I swore a blue streak. The Recon was carrying M-16s, a few M-14s and three M-60 machine guns. But many of the Recondos had brought their damned AK-47s, and every time they opened up we couldn't tell if it was our people or the VC.

We'd moved about 200 meters when Hung got a call on his radio. There was at least a full battalion of Viet Cong on the southern edge of Phu Cuong, less than a kilometer from where we were. There was heavy fighting as elements of the ARVN 1st Cavalry Squadron pushed north into the city. We were ordered to set up blocking positions across the main street (Highway 13) to prevent the VC from pulling back into the downtown area. This probably looked good to the officers back at Division, but with only eighty men, we were going to have a hell of a time putting up an effective block. A Viet Cong battalion usually had between 300 and 600 men, and even after many casualties, this one probably outnumbered us three or four to one. I figured the VC would hit us head-on with at least a company and then, discovering what a small force we were, probably would flank our position or infiltrate around us. I wasn't particularly worried that they would wipe us out because the ARVN 1st Cav's tanks and armored personnel carriers were hot on their heels. If they didn't get through our line on the first attempt, they'd just go around us. Then we'd have to chase after them.

I looked at my watch and saw that it was 1300. I realized that I was hungry; it had been almost eight hours since I had eaten anything. Everyone else must have felt the same way because they were pulling rice balls and bread from their packs. Although we could hear firing to our south, it would be a while before Charlie made his appearance. While we were eating, a Recon NCO came up and said something to Hung. Hung smiled and called to me that his wife and children were okay. He had sent a squad to check on them, and the troops found them hiding in the bedroom. She and the kids had been escorted to the province headquarters compound and were now safe.

A few minutes later one of our outposts alerted us that a friendly battalion was coming up the road from the rear. Somebody had decided that eighty men were not going to be enough and had sent reinforcements. I turned around and watched the troops moving toward us in a double file on either side of the road. Then I spotted their command group marching right up the middle of the street, as if they were on some sort of training exercise. Hung and I looked at each other and shook our heads in disgust. It was pretty stupid to walk down the middle of a street any time, but it was particularly dumb when you knew that there were Viet Cong just ahead.

As the advisors approached, I recognized the officer. He was a new lieutenant who had just arrived in Vietnam. I had talked to him a couple of days before when he had asked me what it was like in the bush. He had looked so young and inexperienced that I'd wondered how he was going to react to his first battle, now just a few minutes away. I yelled, "You'd better get your ass off that road before Charlie blows you away. The fuckers are just up the road."

He smiled, waved, and continued right up the middle of the god-damned street. I looked back at my own people and they shrugged.

As we watched the battalion disappear around a bend in the road, Colonel Quan and Major Stefaniw drove up in a couple of jeeps. Hung got up and went to meet Quan, and I walked over to Stefaniw. Major Stefaniw explained that the VC battalion was on the edge of town and that the ARVN 3d Battalion, 9th Regiment, and most of the ARVN 1st Armored Cav were attacking from the south. The ARVN 1st Battalion, 7th Regiment (the unit that had just walked by) was moving forward to hit the VC from the north. Six or seven tanks were en route to reinforce the 1/7 attack and would be coming up from our rear at any time.

"What are we supposed to do?"

"I think that your unit is in reserve."

"What the hell do you mean, 'in reserve'?! My troops are the best in the entire division, and we've been screwing around since this morning. If we don't get a decent mission soon, I'm going to call Division and tell them that we're going off to find our own war."

"Okay, I'll talk to Colonel Quan, but I think he figures he has enough of his own troops without using outside help."

I walked back to the ditch and sat down. Pretty soon the jeeps drove away and Hung walked over, frowning. He was not the type to sit on his ass while there was a fight going on. I suggested that he call Division and tell the people there that we were being wasted, but he shook his head. "Colonel Quan will need us soon."

I had just finished eating my rice ball when firing broke out up the street. Because of the bend in the road, we couldn't see what was happening, but it didn't take many smarts to figure out that the 1/7 had just walked into the VC. Hung grabbed the radio handset and held it to his ear. After a couple of seconds he stood up and said, "We fight now."

The company began to move forward, and when we rounded the bend in the road, we spotted Quan's and Stefaniw's jeeps next to a brick building. Hung and I double-timed up to them and, in one of the quickest briefings we ever had, were told that the 1/7 was withdrawing and that we were to hold the line against a VC attack.

I asked Major Stefaniw, "What the hell is happening? If the VC are forcing a whole ARVN battalion to withdraw, how does Quan figure that eighty Recon troops are going to stop them? Why doesn't he re-form the 1st of the 7th right here? With both units we can hold Charlie."

Before I could say any more, Quan's jeep took off to the rear and Stefaniw had to jump in his jeep to follow. Almost immediately, panic-stricken soldiers from the 1/7 came running through our line. Some of them were yelling that the VC were outflanking us on the right. That's all we needed—a friendly battalion breaking and running *and* the VC beginning to surround us.

Our troops tried to stop the 1/7 soldiers, but none were about to stay and fight. They just wanted to get away as quickly as possible. In the midst of the mob, I spotted the American sergeant who had passed by just a few minutes before. He had tears running down his face and appeared on the verge of losing control of himself. "They killed the lieutenant!"

I yelled at him, "Are you sure he's dead?"

"They caught him with a burst of machine gun fire. When he went down, they shot him to pieces."

I wanted to make certain that the lieutenant was really dead. Our biggest fear—even greater than being killed—was that we might be captured by the VC. "You're absolutely certain?"

He sobbed. "Yeah—his body was all torn up."

He staggered to the rear, and in about thirty seconds everything was quiet. The last of the 1/7 disappeared around the bend, and we were alone. I looked around and felt pride. My troops were ready—

none had been bitten by the panic bug and they were ready to fight, even the new ones. I didn't see any looks of fear—only disgust.

"What do we do now, Hung?"

"Colonel Quan said that some tanks will be here in a few minutes. When they get here, we go kill some VC." (I liked the way he phrased it.)

A few minutes later we heard the sound of tanks. In a moment the lead tank appeared at the bend behind us. The tank commander was standing in the turret hatch, just like a Vietnamese Patton. I figured that he hadn't heard the fighting because of the tank noise; he probably hadn't seen the fleeing soldiers either. If he had, he would have been buttoned up inside.

Hung went over to the tank, quickly briefed the commander, and then we all began to move slowly forward. The tanks were forced to stay on the street because of the deep ditches and buildings on either side of the road. It was quiet, but we were under no delusions—the VC were waiting. After covering a couple of hundred meters, I saw a pile of bodies in the middle of the road. One of them was the young lieutenant. He'd been killed in his first minute of combat. If the poor bastard had gotten off the road like I'd told him to, he might still be alive. Oh well, I thought, this was not a game for fools or innocents.

The lead tank had just passed the bodies when there was an explosion. I hit the ground, looked to my left, and saw the tank sitting stationary in the middle of the road. The VC had hit it either with a recoilless rifle round or an RPG. The round had hit the forward part of the turret, where it curved up to the commander's hatch. The slope of the armor had deflected the blast up and right into the tank commander. It had hit him at his waist and ripped him wide open. His legs were in the hatch and his upper body was hanging over the back of the turret—head down, face up, and arms dangling. His intestines had been torn free and blown back over his face and head. Blood and half-digested food were dripping on the back deck of the tank—a pretty gruesome sight.

At that moment the driver threw the tank into reverse and began roaring down the road backward. Drivers sit in the very forward part of the tank and can only see to the front. If their hatch is closed, their ability to see is even more restricted. It didn't matter to this driver that he couldn't see where he was going. He only knew one thing—his tank had been hit and the VC were to his front. He wasn't going to wait for them to fire another round.

The second tank was only a few meters to the rear. Before its driver could react, the lead tank smashed into it, knocking it into the ditch along the edge of the road. In seconds the hatches on both tanks were thrown open. The crews piled out and ran back down the road as fast as they could move.

The other three tanks that were farther back shifted immediately into reverse and roared away. I was surprised that they were able to stay on the road at the speed they were travelling, but they all disappeared around the bend, leaving us to fight without any support.

The VC were blasting away at us with everything they had. From the volume of fire, I could tell that we were engaging only part of the battalion, probably a company. The rest of their force was still busy with the two ARVN battalions on the other side of their position.

Then the firing stopped. Firefights are like cocktail parties. Everyone is talking and the room is noisy as hell. Then for some reason, everyone stops and there is a moment of silence. This same phenomenon sometimes happens in battle—except that it's gunfire rather than conversation.

As we lay there catching our breath, a mob of twenty or thirty civilians began boiling out of the houses between us and the VC. They'd been hiding and decided they'd better make a break for safety during the lull in the firing. One minute the 200 meters between us and the VC was a no man's land, and the next it was filled with running civilians.

I was surprised that the VC held their fire. I had expected them to open up, leaving dead and dying civilians all over the place. But nothing happened, and the civilians disappeared down the road behind us. When the last one rounded the bend, both sides opened up again, as if on command.

The shooting continued for a while and then began to slacken off. A man can carry only so much ammunition, and after the initial heavy firing, he has to begin conserving his ammo. It looked as if we had reached a stand-off. Neither side was going to launch a frontal attack, so it was time to figure out alternatives.

I thought about the two tanks sitting in the road. Except for the fact that one needed some new paint, both were perfectly functional and armed with 76mm main guns that could do a job on the VC. Without thinking about how stupid it was, I jumped out of the ditch and dashed across the road to the rear of the lead tank.

Keeping the tank between me and the VC, I caught my breath and began considering what to do next. If I went in through the top, I'd first have to climb up on the back deck and drag the tank commander's body out of the hatch. That shouldn't pose much of a problem—except for the pool of slippery blood and gore—because I could keep down behind the turret. The second part of the exercise would be a lot trickier. I'd have to get up and into the hatch before the VC blew me away. As a few bullets ricocheted off the turret, I decided that going through the top hatch was not a healthy move.

The only other alternative was to crawl under the tank and come up through the escape hatch in the floor. I had been crouching behind one of the rear drive-sprockets and had the hull, tracks, and road-wheels between me and Charlie. When I peered under the tank, I realized that I would have to crawl between the tracks, directly toward the VC's position. If they were paying attention, they would be able to shoot me as I wriggled to the escape hatch.

While I was considering this course of action, I thought about what I would do if I got inside the tank. I'd never had much to do with tanks, so I'd have to figure out how to load, aim, and fire the main gun. It looked so easy in movies; but this wasn't a movie, and I didn't think that Charlie was going to give me much time for "on-the-job" training. When the VC figured somebody was inside the tank, they were going to slam some more RPGs and recoilless rifle fire into it.

I should have thought this all out before I left the protection of my ditch. Now I realized it was really a dumb idea. I looked over to the side of the road and saw that everyone was looking at me. I shrugged my shoulders and made a mad dash back to the ditch, sliding into it like a baseball player coming into home plate. Glen called over to me from the other side of the road. "What's the matter? Decided that it wasn't such a hot idea?"

179

"Shit! It worked so well in that Audie Murphy movie."

As we were lying there, we spotted a whole company from the 1/7 creeping through the buildings behind us. At first we thought they were coming to help, but then we realized they were sneaking away. Hung and I crawled back down the ditch, and he grabbed the company commander. This company had been several streets over and had missed the word to pull back with the rest of its battalion. When they realized they were sitting alone on the flank, they decided it was time to get the hell out. Hung had the ARVN lieutenant behind a house and he was yelling at him to collect his troops and get them back into the line. Then without warning, Hung hauled off and slugged the officer in the face. The young guy fell to his knees, holding his jaw. Hung spit on him and then turned around and headed back to our troops. The ARVN lieutenant got to his feet and staggered off to the rear.

When I got back to my original spot in the ditch, I found that Glen and Tony were gone. I supposed they had moved over to the right, where the Recondos were. I called to Hung. "What do we do now?"

Hung grinned or grimaced—I couldn't tell which—and pointed toward the VC. Those of us near the road began to crawl forward, while the others darted from one building to the next. The VC increased their fire, but most of their rounds were passing over our heads. They were holding slightly higher ground than we were and weren't compensating for the sloping terrain. The air was filled with bullets, but from the dust being kicked up by our rounds, I could see that the Recon's fire was the more accurate.

Chuck crawled alongside and handed me the radio mike. Glen was calling. "We've got a little problem. Your Five [Tony] and I are pinned down by a machine gun, and we need some help to break contact."

"Roger. I'm trying to get the tanks up. Can you use any artillery?"

"Negative. We're too close for artillery to do us any good. We've gotta have those tanks."

"Do you have any LAWs you could use on that gun?"

"Negative. We're cut off from the rest of the people."

"How in hell did you do that, you dumb shits?!"

180

"Sorry 'bout that, but chew our asses *after* we get out of this. Right now we're in a bind."

"Roger. Keep your shit together, and I'll try to get something over to you."

I yelled to Hung that Tony and Glen were pinned down and that I was going back around the bend to try to get the tanks to come up. Since our attack was stalled, Hung yelled that he would go with me. Using the buildings for cover, we made our way back along the road. When we got around the bend, we spotted them—three big mothers sitting in the street with their engines idling. As Hung argued with the Vietnamese commander, I climbed up on the deck to talk to the advisor.

."Hey, man, I got two Americans pinned down by a VC machine gun. Your tanks are supposed to be playing this game, too. How about getting your counterpart to move his ass up where he belongs?"

For the next two or three minutes, the tank commander was catching hell from both Hung and his advisor, but he was terrified. He'd seen what had happened to the guy in the lead tank and had lost his courage. I became so angry that I raised my rifle and pointed it at his head. "Lieutenant, you get these tanks up to the battle or I'll blow your head off."

He began to shake, but he made no move to do what I'd ordered. I clicked my M-16s safety off and had every intention of splattering his brains all over the tank. Suddenly the advisor yelled, "Don't do it! If you shoot him, the army will have your ass. He ain't worth it."

Hung saw that I was serious and put his hand on my arm. "Don't kill him. I will take care of him later."

Just then the VC opened up with a heavy machine gun, and rounds began to hit around us. Without hesitation, all three tanks began firing in Charlie's direction with their .50 caliber machine guns and 76mm main guns. They weren't aiming—they were just shooting in the general direction—so some of their fire was going into the Recon's positions. I raised my rifle to shoot the tank commander, but he instantly dropped down inside the turret and slammed the hatch shut. I jumped off and ran over to Hung. "We'd better get back to the troops."

By the time Hung and I got back, the tanks had stopped shooting

181

and the VC's fire had slowed to just a few rounds a minute. This was getting very frustrating. We didn't have enough troops to attack, and the VC were caught between us and the ARVN units coming up from the south.

Chuck passed me the radio handset and said that Tony was calling. "Five Alfa and I got free, but he took a round in the leg. It ain't too bad, but he says it smarts."

"Can he walk?"

"Roger. He's a little gimpy, but he can move."

"You assholes get back here *now*! Do you copy?"

"Wilco."

Calling to Hung, I said, "Tony and Glen got out. You think its time to put some artillery on those bastards?"

"I've been trying to get clearance, but Division doesn't want to fire. They're afraid that some of the rounds will come down on the friendlies on the other side of the VC."

"Okay, what are we supposed to do?"

"We're supposed to hold here until they decide."

So we lay there, waiting for something to happen. Both sides had virtually stopped shooting, and we were in a stalemate.

Glen and Tony finally made their way back and joined us. I checked Glen's wound and found that the round had taken a chunk out of his shin. The bone might be cracked, but it wasn't broken. It was a relatively minor, but painful wound. When I chewed them out for going off and playing cowboys, they looked like naughty boys, but I knew that they'd do it again at the first opportunity. I reminded them that they were supposed to be advising the Recondos, not fighting their own private war.

I had a good position in the ditch. Although it ran directly toward the VC, I was behind a driveway that crossed it. There was a small culvert to allow the water to flow along the ditch, but I had about six feet of dirt between Charlie and me. There was tall grass on either side that gave me some concealment. Since we couldn't move forward and wouldn't pull back, we just had to wait until we got reinforcements or new orders.

There wasn't anything else to do, so I decided to play sniper again. I used my rifle scope to scan the buildings where the VC were located.

As I was looking through the scope, a Viet Cong suddenly appeared—right in my sight. He was creeping forward along the edge of a building about 150 meters directly to my front. I squeezed the trigger, and when I looked again, I could see the soles of his Ho Chi Minh sandals. I had hit him in the chest and knocked him on his ass. Although his feet twitched, he didn't move. I called across to Glen. "Did you see that? I got that son of a bitch right in the chest."

"Yeah, real John Wayne shit. Bet you couldn't have hit him if you didn't have that fancy scope."

"Hell, I'm just good."

I was searching for another target when Hung called over to me and said that Majors Tu and Clark were at the bend in the road behind us. They wanted us to come back and meet with them. Once more we worked our way along the buildings. This shit was getting real old.

They were sitting in a jeep near the still stationary tanks. We shook hands and began complaining about the worthless tanks and the gutless 1st Battalion. In English, Major Tu asked, "What do you recommend we do?"

"If we could get those tanks moving, we could get into the VC. If we screw around much longer, it's going to be too late, and the VC will escape after dark."

Hung nodded in agreement and Major Tu said, "Okay, I'll get the tanks up with your soldiers."

Major Tu turned around and headed across the road to talk to the commander of the tanks. Clark stayed with us and said, "General Thuan has put Tu in charge, and I think he can get this show on the road."

We stood there for a few minutes and then heard the tanks begin to move. Major Tu came back and said that he'd gotten clearance to use artillery and that it was ready to fire when we wanted it. Our troops were too close to the VC to put the artillery where it would do the most good, so Hung radioed his executive officer to pull the company back to the bend.

As soon as they arrived, the artillery began to fire, and for about fifteen minutes it crashed into the VC's positions. We used the time to reorganize, resupply ammunition, and get our dead and wounded loaded into vehicles for evacuation. When we were ready, Hung gave the command, and we began to move back toward the VC. I watched the lead tank, expecting it to retreat suddenly again, but it didn't; it moved slowly forward, firing both the coaxial and main guns.

The attack was moving along pretty well, and we were closing on the VC's positions when suddenly we began taking fire from both flanks. The VC were attempting to counterattack on either side. Our troops used the buildings as cover and returned the fire. We were getting used to this kind of fighting. We didn't like it, but we weren't surprised.

Again the ARVN tanks panicked and began spraying both sides of the road with machine gun fire. Unfortunately, our people were between the tanks and the VC and were catching hell. I saw one of my troops hit by "friendly fire," and the .50-caliber slug tore a big hole in him. We were all scrambling for cover when an RPG rocket hit one of the tanks. A crewman and the same chickenshit commander who I had almost shot dove out of the tank and began crawling to the rear. The crewman made it, but the tank commander was hit by a bullet and rolled into a ditch. I had a feeling that the round that hit him wasn't from a VC rifle.

I lost track of how long we were at it, but after more heavy fighting we stopped the counterattack and the VC pulled back. When I next looked at my watch it was almost 1800. It would be dark soon, and we weren't in a very good position. The VC couldn't escape to the south because they faced two battalions in that direction. They couldn't get out to the west because of the Saigon River, and ARVN units were moving in from the east. Their best bet was to wait until dark, then try to overrun our small force and escape through us into the city. I wasn't very happy with our prospects.

I yelled at Clark and asked him if he was going to spend the night with us. He gave me a weak look but didn't say anything. Major Tu low-crawled over and said that General Thuan had ordered us to withdraw and return to defend Lam Son. Thuan was worried the VC might hit his headquarters that night. I had to admit that I was not at all unhappy to hear that. We were so badly outnumbered that if we stayed where we were, without reinforcements, we'd be in for a hairy night.

The orders were issued to the platoons and we began to pull back. After a few hundred meters, we had broken contact and were able to move rapidly back to near the police station. We loaded our dead and wounded into the trucks Division had sent—we'd lost five killed outright and about fifteen wounded. The 1/7 and the tanks had lost between fifteen and twenty men.

There weren't enough trucks to carry the troops, so we had to walk the three kilometers back to Lam Son. By the time we got there, it was dark. Hung and I compared notes and concluded that the whole damned day had been a fiasco. Our people had done well, but the tanks and the 1/7 had been worthless. Had they fought even marginally, we would have killed a lot of VC. As it was, we thought we had hurt them, but we had no idea how many we had killed. We'd have to go after them again the next day.

Early the next morning we moved out again—once more on our way to Phu Cuong. Before we left, I stopped by the TOC to find out the current situation. The ARVN troops had finally retaken the engineer compound, and fighting in the rest of the city had died down. No contacts had been reported after midnight, and it appeared that the VC had pulled out.

Hung and the troops looked tired. After fighting for most of the previous day, Recon Company still had to man part of Lam Son's perimeter defenses all night. It was done in shifts, but the most sleep anyone had been able to get was only about four hours. The others in the team—myself included—were also tired, because we had spent much of the night monitoring the action. If we were committed, I wanted to know as much as possible about the situation before going in.

The medics in Lam Son's dispensary had cleaned and dressed Glen's wound and had ordered him to stay in bed for a few days. When I went to wake the others, Glen insisted that he was going with us, but when he tried to walk, he found that his leg was so stiff he could hardly move it. I told him that I wasn't going to be screwing around with a cripple, so he was to stay in the compound.

As we moved down the road, I noticed about half our wounded had been treated and returned to duty. Had we not been so short-handed, we wouldn't have allowed many of them to go. But until we got some of our absent people back, we needed all the rifles we could get. Every platoon had men with bandages, and we all looked pretty dirty and grimy.

We retraced much of the route we had taken the previous day, searching for the VC or indications of which way they had withdrawn. When we reached the engineer compound, I suggested to Hung that the troops could use a break. We'd walked about four kilometers, and the temperature had risen to near the 100-degree mark. He chuckled, because he knew what I wanted to do, and said that we'd halt for thirty minutes. Telling him that I'd be back by the time he was ready to move again, I took off toward the main building.

I found some engineer-advisors and asked them if they had recovered any VC bodies from the building's second floor. They had, but there had been more blood pools than bodies—meaning the VC had been able to get away with some of their dead and wounded. When I asked if any of the VC had been hit by M-16 fire, they couldn't tell me. After playing those stupid games with the tanks, the ARVN finally decided to open up on the building. All the bodies had been so chewed up that it wasn't possible to tell if any had M-16 holes in them. I was a little frustrated, because although I was sure I'd killed several VC here and in the afternoon battle, I could only confirm the one that I'd shot next to the wooden house.

When I got back to the company, Hung gave the word to move. He asked me how many I had killed the day before. I smiled and shrugged my shoulders. "Couldn't tell, because they fucked up the bodies with their goddamn .50 cals."

We'd wanted to go back to where we had fought the day before, but Division insisted that we sweep the northern edge of the city. They had parts of four different ARVN battalions in the south and didn't need us. There was little damage in this part of Phu Cuong, and the people were returning to their hootches. I suspected everything would be back to normal by the next day.

About a kilometer north of the engineer compound, we entered a village that had a couple of hootches hit by airstrikes. I didn't see any bomb craters, so apparently the aircraft had used only napalm. The villagers were standing around and looking at the remains of the two burned homes, and Hung asked one old man what had happened. The guy said that a VC squad had come into the village two nights before and set up a mortar, apparently firing on the engineer compound or the province headquarters. Evidently a Bird Dog or a helicopter had spotted the muzzle flashes and called in an airstrike. I guess the planes had been able to knock out the mortar with napalm, allowing them to save their GP (general purpose, high explosive) bombs for another target.

I saw only two bodies. One was face down at the edge of the burned rubble, and the other was a few meters away, lying on his back. Both were burned to a blackened crisp and covered with flies. The stench of charred flesh was overwhelming. The old man said that they were VC, which explained why the villagers had not buried them. I guess that the peasants were waiting for ARVN troops to give them the okay. No one was weeping or wailing, and I didn't detect any animosity toward us, so I assumed these people must be pro-government—or at least neutral.

By late afternoon we had made a large circle and were headed back to Phu Cuong. Other than the two bodies, we hadn't seen any signs of the Viet Cong. The people we had talked to either knew nothing or pointed to the northwest, indicating that the VC had withdrawn in that direction.

Division sent trucks to pick us up and take us back to Lam Son. General Thuan was taking no chances of having his reaction force out of Lam Son during the night. He and his staff were worried stiff that the VC might try to repeat their aborted attack on division headquarters. I hoped that he would relax soon, because this old shit of breaking off an operation to protect a bunch of REMFs wasn't cutting it.

We weren't the only combat unit in Lam Son. After the first night General Thuan ordered the ARVN 1st Cavalry Squadron to move into the division compound. Its tanks and armored personnel carriers were positioned at intervals all around the perimeter, while the division's support personnel manned the bunkers along the berm. The Recon Company was taken off the perimeter and put in reserve so it could reinforce any threatened sector or counterattack if the VC got through the wire and over the berm.

On the third day of what the newspapers were calling the "Tet Offensive," General Thuan decided that he would combine the Recon Company and the 1st Cav into a task force. Neither Hung nor I liked that idea, and we complained to our superiors. After our experience with the tanks on the first day, we wanted nothing to do with them. As far as we were concerned, they weren't worth a damn, and if they panicked, they could hurt us more than the VC could.

Colonel Sonstelie argued that it was an ideal setup. With the armored vehicles we could move fast, and between our aggressive Nungs and the cav's firepower, the task force could put a real hurt on the Viet Cong. Despite our arguments the issue was academic, because General Thuan wanted the task force and he was going to get it.

Shortly after first-light, the tanks and tracks pulled out of the perimeter, and after forming up at Lam Son's front gate, the Recon and Recondos climbed on board. We had an interesting situation, because while the squadron commander was a major and Hung was only a captain, Hung was given overall command. Technically, the senior officer is normally the commander, but that legalistic bullshit was thrown out the window when General Thuan decided that Hung was the man. A similar situation existed on the advisor side, because the squadron senior advisor was a captain and I was a first lieutenant. However, here there wasn't a real problem, because the Vietnamese were in command and we Americans took our precedence according to our counterparts' positions. Captain Egan was a sharp guy, about my age, but he had less time in the field. He and I hit it off right away when he confided that he hoped some of the Recon's professionalism would rub off on his troops.

When everyone was ready, the drivers started their engines and the convoy lurched out the gate and turned left onto the road. This time we weren't going into Phu Cuong. We were going to make a wide sweep to the east and then turn southwest to the village of Bung, which was six kilometers from Phu Cuong. If we didn't make contact,

189

we would turn northwest and follow Highway 13 back to the province capital.

Because it was the dry season and the terrain was fairly open, this area was excellent for the tanks, but poor for the VC. Except for a thin patch of woods, there was no place for a large Viet Cong force to hide. The rumbling of the armored vehicles gave plenty of warning, so that small VC units could scramble down into their tunnels.

We dismounted before we got to the woods and advanced through the trees on line. The squadron commander wanted the Recon out front to protect the vehicles from antitank weapons, but Hung told him to piss off. He weren't going to risk having the bastards open up behind us. At the far edge of the woods, we climbed back on top of the tanks and tracks. We hadn't found anything and were on our way toward Bung.

Bung, one of the larger villages in this area, was situated where several roads intersected Highway 13. There hadn't been any fighting reported at Bung, although there had been some at Lai Thieu, four kilometers to the south, where the VC hit the district headquarters. Bung was a major marketplace for the surrounding area, and when we arrived, it was crawling with people. You'd never believe there was a war on, let alone a major Viet Cong and NVA offensive. The column stopped long enough for the soldiers to buy bread and something to drink, and then we were on our way again, up Highway 13.

Although the trip was a little scary—we knew that whole VC and NVA battalions were roaming around—it was uneventful until we reached Phu Van. The three lead tanks had just passed through a cut in the road when a Charlie opened up on us with an AK-47. A fraction of a second later, several other AKs joined the first one, spraying rounds at the column. In an instant we all rolled off the tracks and dropped the seven feet to the ground. The fall knocked the wind out of me, but I was able to scramble around to the opposite side of the track, putting it between me and the VC. (We rode on the top of the M-113 because we sure as hell didn't want to be inside if it hit a mine.)

The troops hit the ditches along the side of the road and began firing toward the VC. The tracks opened up with their .50- and .30-

caliber machine guns, and the tanks began firing everything they had—both machine guns and main guns—plastering the side of the road.

From the initial VC burst of fire, I guessed there couldn't be more than a squad confronting us. There might be more, but if there were, they hadn't done anything yet. Since the armored vehicles were putting out so much fire, Hung passed the word for our people to stop shooting. It looked like a case of overkill, and there was no reason for us to waste ammunition—we might need it later.

After a few minutes the firing died down and finally stopped. Hung made certain that the squadron commander radioed his people to hold their fire, and then he gave the signal for the Recon to move into the field. The area was pretty bare, with just a few bushes and some knee-high grass. I couldn't figure out why the VC had picked it as the spot to ambush us. There was no place to hide and no way to escape.

We'd only moved about fifty meters when I heard shouting off to my left—our people had found something. When I got there, several troops were crouching over three bodies dressed in black pajamas. All three were wearing Chicom ammunition pouches, the kind that held AK-47 magazines on the chest. They also had NVA pistol belts, canteens, and hand grenades. The bodies were thoroughly mangled, and their AK-47s were all shot up and worthless as trading material.

We searched the area around the bodies and found an electrical wire and a dry-cell battery. Tracing the wire back to the road, we discovered a 250-pound aircraft bomb that had been rigged as an *anti-everything* mine. If these guys hadn't been such amateurs, they'd have blown the hell out of our convoy. I don't think that they were expecting a whole column of armor. One of them probably panicked when we came through the cut in the hill and opened up with his AK-47. If they had waited a couple of seconds and set off the bomb, it would have destroyed at least one or two tanks or tracks and done a job on the troops riding on the top.

After some nervous grinning and self-congratulations on our luck, we decided to move directly across country, toward Phu Cuong. This would take us right through the battleground where we had fought

191

the VC battalion on the first day. When Hung announced that we would move on foot, with the vehicles following, everyone was relieved.

We'd gone less than a kilometer when we reached the area. The bulk of the VC battalion had been forced to take up positions in a barren, treeless field at the edge of town. The main position was on a slight rise, crisscrossed by several deep ditches and gullies. This gave the Viet Cong good cover from ground fire—but very little from the air. It would have been a stand-off if our people hadn't had artillery and air support. Because they did, it was a massacre. The ARVN units just sat back and pounded the VC.

It pissed me off that we hadn't been told exactly what the situation was. Had we known how bad off the VC were, we would have just dug in and made certain that none of them escaped through the city. But nobody bothered to keep us informed, so we had repeatedly attacked, losing three tanks and a bunch of troops.

My anger was forgotten as I walked around the field—there were dead VC everywhere. Every time I tried to count the bodies, I was interrupted by someone shouting to come see what he had found. The whole area was littered with packs, bandages, ammunition, and equipment. The weapons had been collected by another ARVN unit the day before, but we did find an 82mm Chicom mortar that had been so damaged that it was useless.

I counted forty dead before I finally gave up. The other units' advisors would certainly have made a count, so I was probably wasting my time. This was the second full day after the battle, and the hot sun was doing its work on the dead. The bodies were badly bloated and the smell was awful.

Some of the dead were lying in the bottom of the ditches, but many were curled up in shallow caves that the soldiers had tried to scrape out of the sides of the gullies. It must have been pure hell, and I hoped that the VC commander had been killed. He deserved it, because he got his unit into an impossible situation.

There were a lot of dried pools of blood, bandages, and empty holes, which told me that many VC had escaped after dark, when the Recon Company had been ordered to withdraw. Since the VC made every effort to recover their dead, forty-plus bodies indicated a very large force, with many casualties. I estimated that it probably had been a reinforced battalion—and possibly more.*

As we gathered up our people to move out, I was struck by an odd fact: No napalm had been used. This didn't make any sense. If the air force had put napalm on such an exposed position, it would have really done a number on the VC. I doubt that there would have been any survivors. Either they didn't use napalm because they didn't want to start fires that might have spread into Phu Cuong, or they were experiencing another of the ammunition shortages that plagued our forces from time to time. Whatever the reason, it meant that we would have to meet this unit again after it recovered from its losses.

The next day the Recon went out alone. The tanks and tracks had taken a beating and needed maintenance and repairs. We could have used some maintenance ourselves, but all we got was a little sympathy and new orders. We were, however, getting some of our troops back from leave. As soon as they could travel the roads, they began making their way back to Lam Son. Many, particularly the ones who had gone to Saigon, told hair-raising stories of evading the VC while trying to get back to the unit. We were still badly understrength, but now we could field a little over a hundred men. Hopefully, in a few days we'd have them all back.

We climbed into the trucks and headed down the road to Phu Cuong, where we unloaded in front of province headquarters. Division wanted us to search along the east bank of the Saigon River and south toward where the smaller Ba Lua River joined it. A battalion of the 7th Regiment was working west along the Ba Lua, from Highway 13 to the Saigon River. If there were any VC in the area, we hoped to catch them between the two units.

As we moved down the road, I checked my map and found that

---

*This had been an NVA battalion with no relatives to recover the bodies. They just lay in the field and rotted. In a few days the stink was drifting over Lam Son, two kilometers away. The peasants living nearby merely ignored the smell and went about their business as usual. How they coped with the odor was beyond me. Finally, after a week or so, General Thuan ordered some soldiers to scatter lime over the bodies, but nobody buried them. A month later we passed through the area and found bleached skeletons lying everywhere.

we were on the same street that Charlie had used for his parade, the one that Hooker and I had seen from the airplane. An hour later I saw two bamboo poles and a white rag lying alongside the road. It was the banner that I had seen while flying over the parade. It was all wadded up, but it was still in good condition. I unrolled it and spread it out on the ground. In carefully printed red Vietnamese letters, it said:

To all the citizens. This is the morning to celebrate a new season of life. Great Victory!
   The National Liberation Front of South Vietnam.

Lying next to the road were several long strings of small Viet Cong flags, which were made out of construction paper. The VC had gone to a lot of trouble making decorations for their parade, but apparently they had run short of paper, because the bottom half of some Viet Cong flags was green rather than blue. I rolled up the banner and stuffed it into my rucksack, figuring it might come in handy as trading material. We piled up the rest of the decorations and left them burning in the middle of the road.

An hour or so later, Hung got a call on the radio, saying that the ARVN battalion to our south had run into at least a reinforced VC platoon. (We'd already guessed that, because we had heard M-79 explosions and automatic weapons fire.) Hung gave a few commands and we began quickly moving toward the firefight.

Because of the dry ground, we were able to make good time. Within a few minutes we began to hear a heavy volume of fire, not more than a few hundred meters away. Hung halted the company and talked to the battalion commander over the radio to find out what the situation was. We didn't want to go charging into a firefight without first coordinating with the other unit.

Leaving the bulk of the company in place, Hung and I took a platoon and moved to meet the battalion commander. We found him and his command group in a ditch alongside the road. While Hung crouched with the major, I talked to the battalion advisor, a captain whom I knew only by name. The battalion had the VC pinned against the Saigon River. The ARVNs had Charlie inside a large U—a company was on each side, and one was across the base of the U. There was little chance that the VC could escape across the river, because it was wide and fast-moving at that point, but the battalion commander had called for gunships to make sure. It was an ideal situation—except the battalion was reluctant to assault the VC.

Hung got up, climbed out of the ditch, and walked over to where I was squatting. He laid his map on the road and showed me where the ARVN and VC positions were. He had arranged with the battalion commander for the Recon to take over the fight. The ARVN companies on the north and south would continue to block while we moved through the company on the east to clean out the VC. The American captain looked a little surprised and suggested that it was a dumb idea. "Why not just use artillery to blast Charlie?"

"Well, if we put artillery in there, the VC would just move in close to your people, where the artillery couldn't fire. We'd all sit here until it got dark, and then Charlie would either slip away or break through your lines."

Hung had radioed for the rest of the company to come up, and its men were just appearing around the bend. After a quick briefing, we left the road and started moving toward the company that was holding the base of the U. The area between us and the river was fairly open, with waist-high grass and random clumps of bushes. There were quite a few tall palm trees, but they were scattered enough to allow us to see a hundred meters or so. We could see two or three thatched hootches and one fairly substantial wooden house off in the distance, near the river.

Both Hung and I were as concerned about the ARVN companies on our flanks as we were about the VC platoon. We were both constantly talking to the battalion command group to make certain that their people didn't start shooting into ours. We had just passed through their lines when the VC began firing again. The air was filled with bullets cracking by and thumping into the trees around us. Keeping low, we darted from behind one tree to the cover of the next. Just as they had been trained, some soldiers would fire around trees while the others crawled or dashed forward to the next cover. The army calls this "fire and movement," and the Recon was doing it as well as any training demonstration I'd ever seen at Fort Benning.

The closer we got to the VC, the more frantic their firing became. They knew that they were in a trick and obviously hadn't expected this kind of attack. They probably figured that we'd do exactly what the American captain had recommended. I would have loved to have seen the looks on their faces when they realized that we were actually going to assault *into* them.

We had been firing M-79s into their positions, but now we were close enough to use hand grenades—it was about to get serious. The sounds of the battle were so loud that I could barely hear the squad and platoon leaders shouting commands to their people. Chuck and I were having problems, because every time I made a dash and got to the end of the cord, the radio handset would jerk out of my hand. I was trying to coordinate with Tony and the Recondos, but I finally had to give up. There wasn't enough cover, and the trees weren't big enough for both Chuck and me to move together, so I yelled over my shoulder for him to just follow me.

From the sounds of the AK-47s, it was clear the VC were pulling back and concentrating around the wooden house I had seen earlier. I could see that there was a machine gun firing at us from one of the windows, so I paused behind a tree and pulled off the M-72 LAW that I'd been carrying. I snapped off the end-caps and extended it to its firing position. After checking to see that Chuck was clear of the back-blast area, I estimated the range and put the sight's fifty meter mark on the window. I squeezed the trigger lever and the rocket roared toward the house, exploding in the ground just to the right of the window. I'd jerked when I squeezed the trigger, and the rocket had just missed the target. Embarrassed, I looked around to find Chuck grinning and shaking his head from side to side.

A couple of Recon troops also fired LAWs at the house, but they were even wider of the mark than I had been. Although all three of us had missed, the VC machine gun crew knew that it was just a matter of time before somebody got lucky and blew their asses away. The next thing I saw in the window was a white rag waving at the end of a stick. Somebody yelled at the house and an RPD machine gun came sailing out the window. A few seconds later the door slowly opened and a wounded VC crawled out with his hands in the air.

After some more shouting, four more VC cautiously came through the door with their hands up.

The loss of the house and machine gun was the final blow, and VC began surrendering. The first couple were shot as they rose out of the bushes, but then we realized that they were giving themselves up. There were no diehards in this platoon, and in a few moments all firing ceased.

I stood behind the tree, covering the machine gun crew as they got to their feet and walked toward us with their hands high over their heads. When they reached my tree, a couple of Recon soldiers ordered them to lie on the ground. Then the soldiers searched them and tied their hands behind their backs. When this was done, they were ordered to stand up.

Son of a bitch! One of them had been a cook for my old battalion! This fucker had cooked all the meals for our command group during the seven months that I was with the battalion. I grabbed the guy and asked him how long he had been a Viet Cong. Vigorously shaking his head, he replied that he wasn't VC. Just then Hung came over and asked what was going on. I told him that the prisoner had been my cook when I was with the 3/7. Hung smiled and said, "A VC spy."

"No, I don't think so. This guy was a cook because he was too dumb to be a rifleman."

I didn't know the Vietnamese word for "retarded," so it took me several minutes of trying to find an alternative word before Hung finally understood what I meant. I questioned the man some more, and he told me that he had been visiting his family in Saigon when the VC grabbed him and drafted him into their army. He hadn't been wearing his uniform, and he didn't tell them that he was already an ARVN soldier for fear they'd kill him. He said that they hadn't given him a weapon; instead, they made him carry ammunition and equipment. He was carrying ammo for the machine gun when this fight started.

I checked the five prisoners from the house, and all but two had red marks on their shoulders from firing weapons. The cook and one other man had strap marks on their backs, but nothing on their shoulders. I wasn't being very scientific, but since the house had netted only two AK-47s, I concluded that the cook's story could be true. Besides, I didn't think that he was sophisticated enough to have passed himself off as mentally retarded for as long as he had been with the 3/7.

I told Hung my conclusions and asked him to release the man. He said that I was crazy, but he agreed to turn him over to Division when we got back and to tell them the circumstances under which we had taken him. He also would tell them that I thought the man's story was true. That was fair, so I told the man what we had decided. I also told him that we would track him down and cut off his balls if we ever discovered that he really was a VC. *

We moved back to the road and turned over ten prisoners to the ARVN battalion commander. We told him that there were another seventeen dead bodies lying in the field and that he could do whatever he wanted with them. We called in a Dust Off to evacuate the five guys who'd been wounded. Then we radioed for a truck to bring us an ammunition resupply and to take the captured weapons and our two dead soldiers back to Lam Son.

It was still early, so we turned northeast and worked our way back toward Highway 13. We didn't run into anything, but we spent a very unpleasant two hours waiting for the trucks; the breeze was blowing in from Phu Cuong, bringing with it the smell of the rotting bodies lying at the edge of town.

That night General Thuan relaxed a little and let the Recon stay in its own compound across the road from division headquarters. It was a good thing, because at 2300 we heard several big explosions between Lam Son and Phu Loi. I was curled up on the floor in Hung's hootch when I heard the first one, and I rolled into the ditch before the second one hit. (The troops had dug a deep ditch along the edge of the hootches, having cut the lower three feet out of the hootch walls. Whenever the VC mortared the compound, the troops

_____

* A couple of months later I checked and found out he was back cooking for Captain Ninh and the command group. They had agreed with me that his retardation was real, but they made certain to keep him away whenever they discussed operations.

198

could slide across the floor and drop into the ditch without having to run out the front doors and around the buildings.)

Although I was fast through the wall, I wasn't fast enough in the ditch to avoid being knocked flat by Hung and his radio operator as they slid out of the building on top of me. After picking ourselves up, we moved to the right so we could see around the hootch toward Lam Son and Phu Loi. Way off beyond the two camps, we could see 122mm rockets streaking into the sky, trailing flame from their tails. They went up, arced over, and then crashed into the ground with big explosions. Charlie was rocketing the Phu Loi airstrip, but his aim was so bad that the big rockets were coming down all over the place, mostly in the open area between the strip and Lam Son.

The Soviet 122mm rocket is a bad hummer. It's about nine feet long and almost five inches in diameter, and it weighs a little over 100 pounds. It's designed to be fired from a rack mounted on a truck, but Charlie merely propped it up with some bamboo poles and launched it in the general direction of whatever he was aiming at. It was definitely what is called an "area weapon," because it might come down anywhere—but when it did, it made a big bang.

I had been shot at by most of the stuff in Charlie's inventory, but this was the first time that I'd ever been around 122mm rockets. They were normally reserved for REMFs and their big bases. No VC was going to fire an inaccurate 100-pound rocket at an infantry unit, particularly after having humped it all the way down the Ho Chi Minh trail.

This was a new experience for Hung as well, and we wanted to watch the show. We could see where Charlie was launching them from and could trace their flight all the way into Phu Loi. We climbed up on the edge of the trench and sat there through the whole performance. When one of the rocket trails looked like it might come down near our compound, we ducked down into the ditch until the thing exploded. The nearest one blew just across the road, next to General Thuan's TOC. That one sure got the division staff's attention.

Things really got interesting when the 1st Infantry Division's artillery began shooting counterbattery fire. The Americans had some sort of magic radar gizmo that could pick up the flight of the incoming round and plot the location from where it had been fired. Then the information was telephoned to the artillery and they'd poop rounds back at the VC gunners. It must not have worked as well as they would have liked, because soon helicopter gunships began to circle over Charlie's position, firing 2.75-inch rockets. This ended the show for the night, and we climbed out of the trench and went back inside the hootch. We bullshitted about the show for a few minutes and then fell asleep.

The next morning we moved out; our orders were to sweep the rocket launch area. It turned out to be a waste of time. While we found the bamboo poles and the blast marks from the rocket motors, we didn't find any evidence that the counterbattery fire or the gunships had killed anybody. In fact, I didn't find a friendly artillery crater closer than several hundred meters from the launch site.

In late afternoon, after returning from the rocket search, I checked in with the division TOC to get updated on the current situation. Although there was still fighting in Saigon, there wasn't much happening anywhere in our three provinces. I looked over the combat status chart and saw that the strength figures for all of the division's battalions had been constantly climbing; the soldiers who had been on Tet leave were gradually making their way back to their units. The chart also showed friendly casualties and the VC body count. Compared to the number of Viet Cong who had been killed in the last seven days, the division had relatively few casualties, and the "kill ratio" was higher than I had ever seen it. Instead of our having to hunt the VC down and kill a few here and there, Charlie had made the mistake of coming out in the open and trying to fight a more conventional war—and we were waxing his ass.

I wandered over to the aviation section and found its people planning a Firefly mission. A typical Firefly operation involved two helicopter gunships and a Huey equipped with a searchlight. After dark the Huey would fly around looking for VC with a bank of seven powerful aircraft landing lights shining out the right door. When it spotted a target, the gunships would go in and pound the VC with rockets and 7.62mm miniguns.

This particular mission was going to be a bit different, because the aviation people were going to use the new AH-1G Cobra gunship. The old "Charlie" model gunships were utility helicopters that had been modified by hanging various weapons on them—rocket pods, 7.62mm miniguns, 40mm automatic grenade launchers, and whatever else struck the army's fancy. The new AH-1G had been designed from the ground up as an attack helicopter, and it looked the part—sleek and mean.

While I liked the looks of the Cobra, I still preferred the Charlie-model gunship. The first time I used Cobras, the pilots had fired their rockets from way up in the air. When I radioed that I wanted them to fly lower, they responded that they could hit the target from the altitude they were at and that they had been ordered to stay high to reduce their chances of being shot down. The last part didn't go over too well, so I canceled their mission and sent them packing. Then I called Division and asked for a couple of Charlie models that would get down and wrestle with the VC up close—like us grunts did.

Since the Cobras looked so neat, I didn't want to give up on them, particularly after only one mission that may have been flown by pilots with no balls. If I could wangle my way on this Firefly operation, maybe I could get a better look at how the Cobra worked.

As I listened to the aviation liaison officer brief the pilot in charge of the Firefly mission, I made certain to interject a few comments about where his people were likely to find Charlie. After my second interruption, the liaison officer introduced me. "This is Lieutenant Parrish; he is the Recon Company advisor. If anyone in this division knows where Charlie is, he probably does."

With that introduction, I proceeded to overwhelm them with everything that I knew about where the VC liked to operate, but I threw in a lot of "ifs" and "maybes." Finally I stopped and said, "It's always pretty fluid, but I'd be willing to go along with you and show you where we have tracked Charlie."

They thought that was a good idea and accepted my offer. Folding their maps, they walked me out toward the airstrip. As I stood admiring the two Cobras, the pilots went through an abbreviated preflight check and started the engines. I turned and walked around to the right side of the searchlight Huey and found the barrel of a big .50-caliber machine gun sticking out the side—this was going to be okay!

We took off and headed northwest as I guided them to my favorite place, that part of the Saigon River flowing by Phu Hoa Dong and the Hook. When we got over the river, they turned on the searchlight and aimed it at the west bank. The seven landing lights threw a hell of a beam, and they lit up the ground like a stage spotlight as we slowly worked our way north, searching both sides of the river.

Just as we rounded the bend in the river, the light picked up two small sampans hugging the bank. Besides the fact that no boats were allowed to move after dark, there was nothing friendly in this area. On the one side of the river was the Hook, and on the other was the Iron Triangle. Whoever was in those boats, they definitely weren't innocent civilians. Instead of calling in the gunships, the door gunner just opened up with the .50-caliber machine gun and blasted the shit out of the sampans. The whole helicopter shook with the vibration of the big gun as it chewed up the little boats. In short order there was nothing left but bits and pieces of wood and debris. I had seen at least a couple of VC in them when we started, but now there was no sign of life. They hadn't gotten to shore and weren't floating in the water, so I assumed that whatever was left of them was at the bottom of the river.

We circled the spot a few more times and then started back up the river. I hadn't seen the Cobras, but so far we hadn't needed them. I slapped the door gunner on the shoulder and asked him to let me use the .50-cal. He agreed, and we changed places.

Just west of Phu Hoa Dong, our searchlight struck a much bigger sampan; this one had two smaller ones alongside. There were VC in all three boats, and it looked as if they were off-loading weapons and supplies. I didn't have to say anything, because the pilot also spotted them and circled the aircraft to the right, keeping the boats in his beam. I squeezed the trigger and the .50-cal began bucking in my hands. The mount didn't seem as stable as the one I was used to on the armored personnel carrier, but when you're putting out 800 rounds per minute, you can afford to have a few bullets miss.

I concentrated on the bigger sampan, trying my best to chew it to pieces. I was raking it with fire, when suddenly there was a big explosion that threw pieces of the boat high into the air. I'd hit something, probably claymores or TNT. Whatever it was, it blew the hell out of the sampan and set the wreckage afire—scratch one re-supply ship.

The pilot came on the intercom and told me to cease fire. The Cobras wanted to get into the act and were going to go after the smaller sampans. We pulled back out of the way and sat there watching the gunships make rocket and minigun runs on the boats and along the bank of the river. There was another big secondary explosion on shore when one of the rockets set off the ammunition that Charlie had ferried to the bank. It was all over in a few minutes, and we were off hunting again.

We spent another thirty minutes searching, but then we gave up the idea of surprising any more sampans. Between the .50-caliber, the rockets, miniguns, and the searchlight beam, every VC up and down the river now knew what we were up to. The pilot asked me for suggestions on where we might use up the rest of the Cobras' ordnance. I told him to turn off the searchlight so the VC wouldn't see us coming, and I gave him directions to a point about halfway between Ben Cat and the southern tip of the Triangle. When I figured that we were somewhere near where I wanted to be, I told him to switch on the light.

It only took about two minutes to spot something moving in the brush. I had no idea how many there were, but a target is a target, so I opened up with the .50-cal. I was walking the rounds down a stream line when four or five green tracers came streaking up toward the chopper. I had to hang on for dear life as the pilot kicked the chopper over and flew out of range. That sucker had good reflexes.

The VC stopped shooting, but the Cobras had no problem finding the spot because my tracers had started a couple of small fires in the grass. We sat back and watched as the gunships rolled in one by one and fired their rockets. One of the VC either had big balls or a little brain, because every time a Cobra pulled out of a run, a few tracers would come up from the bushes. It didn't look like Charlie was hitting anything, but he was sure giving his position away to the trailing Cobra.

Because the gunships had expended a lot of their ordnance on the sampans, this fight was pretty short-lived. It seemed as if only a few minutes had passed before the pilot said that we were done for the evening and that it was time to go back home. We had no idea how many VC there had been in the stream line or how many we had killed, but we did know that we had fucked up their evening.

When we landed at Lam Son, the mission commander shook my hand and said that I could come with them anytime I wanted. This had been one of their better nights. As I walked back to my hootch, I thought about how these aviators had it knocked. They didn't have to hump rucksacks, push their way through thick jungles, sleep in the mud, trip booby traps, or try to hide behind a water buffalo turd while every VC in the area shot at them. Sometimes they were surprised by a heavy machine gun or a lot of ground fire, but they didn't stumble into L-shaped ambushes and get blasted with claymores and automatic weapons. On the other hand, we grunts made smaller targets and didn't fall as far when we were hit. I'd seen the remains of enough crashed helicopters and crews to know that they weren't on a lark. I really couldn't bad-mouth combat aviators—everything was relative.

Except for the 122mm rocket attack, it had been seven days since the VC had tried anything against Lam Son, but General Thuan was still worried that they were going to do something. Thuan, like many Vietnamese, put great stock in fortune-telling, and he had his personal horoscope prepared each week. According to his astrologer, there was some inauspicious shit in his future, and that was one reason why we were being held on such a short leash. We'd go out in the daytime and then return to Lam Son to provide a rapid-reaction force at night. When I had asked Hung what he thought of astrology, he replied that he thought it was a bunch of bullshit and was keeping us from running up a big VC body count.

On this day, we were sweeping the area south of Lam Son and east of Phu Van. We had done this from time to time in the past and had never found any VC or any sign they'd ever been there. It was a waste of effort, and the Recon thoroughly hated having to go there. That morning, when Hung told us where we were headed, I said, "Oh, you mean the Lam Son Secret Zone."

Puzzled, Hung asked me what I meant.

"You know how every place seems to have a name, like 'War Zone D' or the 'Iron Triangle'?" Well, this one should be called the 'Secret Zone'—because even the VC don't know about it."

The one good thing about the Secret Zone was that we could completely search the area in half a day and get back to Lam Son in time to do some maintenance on our weapons and equipment before it got dark. So in early afternoon, I strolled back into my hootch at Gosney Compound to do some of my own maintenance and repair.

When I walked in the door, everybody started asking me if I had heard about Sergeant Stanowski. One of the captains said that Stanowski was lying wounded in the middle of Tan Thanh Dong and that nobody could get to him. Earlier in the morning, a couple of ARVN soldiers had left the battalion compound and walked into the center of the village to buy some bread. They ran into a large VC force that had somehow infiltrated into Tan Thanh Dong without being spotted by any of the battalion's patrols. One soldier had been killed, and although wounded, the other made it back to the compound to report on what he had seen. Thinking that the soldier was exaggerating about the number of VC, Sergeant Stanowski and the lieutenant-advisor had taken a platoon into the village to check it out. When the VC opened up on them, Stanowski and several ARVNs had been killed or wounded, and the rest of the platoon withdrew back to the compound. The sergeant had been hit in the spine and was lying paralyzed in the open, between two houses. The VC knew the battalion would try to rescue him, so they didn't kill him. They just set themselves up so they could blast anyone trying to get to him. The ARVN battalion had tried several times, but each time the troops were driven back by heavy fire. Now there was a standoff and the ARVN commander couldn't or wouldn't attack again.

I took off out the door and double-timed it over to the TOC, where I found most of the officers huddled around a radio. There didn't seem to be any other activity—certainly none of the kind I would have expected, given the situation. In the hushed silence I heard a weak voice coming over the radio speaker. It was Sergeant Stanowski, and he was obviously in great pain. Over and over he kept asking for someone—anyone—to help him.

Seeing Colonel Sonstelie standing next to the radio, I pushed through the crowd and asked him what was being done to help Stanowski. He told me the same story I had already heard in the hootch—the ARVN battalion commander was saying that his troops couldn't get to the wounded sergeant without taking too many casualties.

"The Recon has just come back from its operation and is still locked and loaded. If you can get some choppers, we can move immediately and save that poor bastard."

"I've been on the horn to III Corps since the very beginning, and they don't have any helicopters they can give us before tomorrow morning. I've also called the 1st and 25th Divisions, but they can't help either. There's no way we can get your Recon to Tan Thanh Dong before first light tomorrow."

Seeing Hooker standing near the door, I went over and said, "Let's get your airplane and see if we can do something."

Hooker didn't even kick the tires before we were taxiing for takeoff. Tan Thanh Dong was only about ten kilometers west of Lam Son, so before I could get my web gear stowed under my feet and my map unfolded in my lap, we were over the village. Hooker called the advisors on the ground and got an update on the situation. The ARVN troops were near the middle of the village, about 800 meters south of the battalion compound. The VC were about 200 meters beyond, and Sergeant Stanowski was lying almost midway between the two forces. Even though Hooker and I were four or five hundred feet above the village, we could see Stanowski sprawled on the ground with a radio on his back.

The advisors were using an alternate radio frequency, because Stanowski had been holding the push-to-talk button down, blocking all other transmissions. He was obviously delirious, because he kept repeating, "Help me," over and over, without listening for a response.

Hooker flew a couple of kilometers north of the village and told me to get ready for a very low pass. Even if we couldn't talk to Stanowski, maybe he would be conscious enough to hear the noise of our engine and know that someone was trying to help him. Screaming in at full throttle, our wheels were only a few feet above the thatched roofs that were whizzing by in a blur. We were moving so fast that the VC didn't have time to shoot at us.

For the next couple of minutes Hooker flew "Lazy Eights" over Stanowski, constantly varying his altitude and approach to throw the VC off. Each time I blasted away at the Viet Cong with my M-16, but I knew that it wasn't very effective. Finally, when we took several AK-47 hits, we knew that we had pressed our luck as far as we could.

As we were circling high overhead, trying to figure out what to do next, Division called and said we were authorized to use artillery if we thought it would do any good. That was a surprise since Tan Thanh Dong was a friendly village—but then Tet had changed a lot of rules. The artillery really wasn't going to be much good; because the VC were so close to Sergeant Stanowski, we couldn't use it right on top of them without hitting him too. On the other hand, if we started it behind them and walked it toward their position, there was a chance that they might get down into the villagers' bunkers; then the ARVNs could sneak in and get Stanowski. It probably wouldn't happen, but anything was worth a try.

Hooker continued to circle while I coordinated what we were planning with the advisors on the ground. Division came back on the radio and said that it had a 1st Infantry Division battery of 105mm guns standing by, and I was given the radio frequency and call sign. I switched one of the radios to the frequency and gave the battery a call. Somebody answered immediately, saying he had been briefed on our situation and was ready to give us whatever support that his people could. I gave him the map coordinates of a point just beyond the southern edge of the village, and I asked him to have his battery fire only its base piece. I didn't want to unload with all six guns until I was certain where the rounds were going to go. I also wanted to give the villagers a warning that some heavy shit was about to happen—as if it were necessary.

Just about when I had the rounds where I wanted them, I got another offer of support. Lam Son had contacted a 25th Infantry Division artillery battery near Cu Chi, and it was willing to fire if I wanted it.

I told them that I was already firing another battery but could use them too. Hooker had a PRC-25 field radio strapped to the back of his seat, so I switched it over to their frequency and gave them some coordinates along the eastern edge of the village. Hooker came on the intercom and suggested that I give him one of the batteries to adjust, but I told him just to keep flying the airplane. I was afraid that unless the whole thing was controlled by a single man, something would screw up, and we would start dropping rounds in the wrong places.

At first everything went okay. I had the switch for the aircraft radio in my left hand and the handset for a PRC-25 in my other hand. I felt like an auctioneer as I'd press one button and say, "Drop five-zero," and then the other one, saying, "Left one-zero-zero, drop five-zero." As the two sets of rounds got closer together, I asked the 25th Division artillery to fire white phosphorous from one of its tubes so I could distinguish its rounds from the 1st Division's.

I brought the twelve tubes of artillery as close to Stanowski as I dared. Then I yelled at Hooker to contact the ground and tell the people there to make a try at getting the man. After a couple of minutes, Hooker came back on the intercom, saying that when the ARVN had tried to move, the VC had opened up on them—the plan hadn't worked.

I kept dropping artillery into the area, but the Viet Cong wouldn't cooperate and the ARVN wouldn't move. There was a lot of bitterness in my voice when I ordered the artillery to cease fire and told them that we hadn't been able to get to the wounded sergeant. I threw the PRC-25 handset to the floor of the airplane and screamed every obscenity I could think of. Technically, the ARVN commander had done the right thing by not risking more casualties trying to save Stanowski, but I didn't buy it. My old ARVN infantry battalion had never left wounded men to the enemy, and my Recon and Recondos would do *anything* to recover their casualties—both wounded and dead. That was one of the big differences between a good unit and a poor one. Good units didn't abandon their people.

It was getting dark when we landed, and Hooker and I walked to the TOC in complete silence. There were still quite a few people in the operations center, but they were just sitting quietly. What little conversation took place was in hushed tones. Stanowski was no longer pleading for help. His radio battery had run low on juice, he had passed out, or he was dead.

At 0700 the next morning, all the Recon and Recondos were sitting along the edge of the runway, waiting for the slicks to arrive. Earlier I had checked with the TOC and found that the VC had pulled out of Tan Thanh Dong during the night. The ARVN battalion was finally sweeping the village, looking for stay-behinds. They found Stanowski dead, with one round in his stomach and another in his head. He was probably already dead when the VC shot him the second time. There was so much blood around him that he couldn't have survived anyway.

The village was secure when we arrived, so we landed on the road in front of the battalion compound. Hung and I climbed out of our bird and walked over where the three advisors and a handful of Vietnamese soldiers were standing next to a row of bodies. All the dead were wrapped in ponchos, but Stanowski was a big man, and his jungle boots extended beyond the edge of the rubberized nylon.

When all our troops were on the ground, we began moving east through the village toward the Saigon River. At the edge of the village, we halted while I got some artillery going into the trees and brush in the distance. We had to cross three kilometers of dry, open rice paddies, and I wasn't going to take any chances. The battery had just fired the fourth adjusting round when there was a big explosion about 100 meters behind me. As the gray smoke boiled up from the hootches, I screamed for the artillery to check-fire—someone in Cu Chi had screwed up and one of the rounds had fallen short.

I raised holy hell with the fire direction center. How in hell could they have been so far off target?! This wasn't some minor error; we were firing on a target several kilometers away, and the first three rounds had been no closer than one and a half kilometers from our position. For a round to fall fifteen hundred meters short meant a major fuck-up.

As I moved back to where the round had exploded, the artilleryman was arguing that he couldn't have been that far off. No one had been hurt, but everyone was pretty shaken up. You don't expect to have a short round that far from the target, particularly when you haven't even started the operation.

I found the crater between two hootches and immediately had an odd feeling. It wasn't as big and deep as it should have been, and the hootches weren't that badly damaged. Then I walked up to the crater, reached down, and picked up a fin assembly from an 82mm Viet Cong mortar round. It hadn't been the artillery—Charlie had fired a single mortar round at us. Apparently I had been getting the artillery close to where the VC were, and the sneaky bastards had pooped a single round back at us to make us think that our artillery had fired short. They knew that we'd stop firing and be disorganized until we figured out what had happened. By then they would have un-assed the target area.

I grabbed the radio handset and quickly explained to the artillery what had happened. Then I asked them to fire a "battery five rounds" on the last coordinates I had given him. Maybe one of the thirty 105mm shells would get the VC mortar crew.

While the artillery was falling, the Recon dog-trotted across the dry paddies to get as close to Charlie as possible before I shifted the fire. I also had Tony get on his radio and ask Division for a light-fire team. Maybe gunships could spot the VC moving through the high grass between us and the river. Two kilometers is a long way to run, so we had to slow to a walk and catch our breath before we were halfway across the paddies. I reduced the artillery to one round every few seconds and finally, when we got close, shifted it farther to the east.

When we got to the initial impact area, the troops cautiously searched it, but they found nothing. Whoever fired the mortar had gotten out of the killing zone while I had been raising hell with the artillery. I called an end-of-mission to the fire direction center and told them that I would let them know if we found anything. I also apologized for chewing their asses.

It took the rest of the day to move up to the river. There was a VC or NVA battalion somewhere out here, and we weren't going to stumble into it by being foolish. Just before sunset we stopped and set up for the night. We put out trip flares and claymores, hung hammocks, and ate our rice and fish—exactly the same things every ARVN unit did before nightfall.

When it was completely dark, Hung put his hand on my shoulder and we got up. Very quietly the whole company moved three hundred meters across the field to another tree line and settled in for the night.

Shortly after 0100 I woke to the sound of mortar rounds exploding in the old position, among our empty hammocks. A few minutes later a line of muzzle flashes began to advance across the field to our front. The mortars, AKs, and machine guns were making so much noise that apparently the VC hadn't noticed that no one was shooting back. As soon as one of our trip flares went off, I knew that they had reached the position and I yelled, "Fire!" into the radio handset. In a few seconds 105mm artillery rounds began exploding on top the VC—right where I had plotted a target the previous evening.

The VC milled around for a few moments and then began running in all directions. I requested artillery illumination, and when it began popping overhead, we opened up with every weapon we had. We didn't get them all, but between the Recon and the artillery, we wiped out a bunch of VC.

There was sporadic shooting during the night as not-yet-dead VC tried to crawl out of the area, but we had to wait until dawn to sweep the area. There were still other VC around somewhere, so we weren't going to take any chances. When it was light enough to see, we sent a platoon forward to search the area, holding the other two platoons in the tree line to cover them. Just as I had thought, the bad guys were NVA, not local Viet Cong. The NVA didn't know us and didn't know that the Recon never spent the night where we ate our evening meal—and that we *never slept in hammocks*. We only carried hammocks to make the VC believe that we'd be in them after dark. We went through this same ruse almost every night, and because of it, we were never mortared in our night positions and never ambushed when we moved out in the mornings. If we had to stay in a position after dark, we made damn certain that we had good cover and could hold off an attack by a large VC force.

While we waited for a chopper to come in and pick up the VC weapons and equipment, I finished a little project that I'd been working on. Using a small stick, I had scraped the caked powder out of a smoke grenade. Then I filled it with C-4 plastic explosive from a claymore. Using a fuse from an M-26 fragmentation grenade and the handle from a smoke grenade, I made it look like a normal smoke grenade—but it was actually a bomb.

As we began moving out, I placed the smoke grenade bomb near one of the VC bodies knowing that Charlie would come back to recover his dead and find it. Whenever we used gunships or airstrikes, the pilot or FAC would ask us to mark our position with smoke. Frequently, when we were close to the VC, Charlie would pop smoke too, creating a lot of confusion as to who was where. Sometimes it would take several tries before we finally got the right positions marked, and the VC often used the delay to move to another location. I figured that when my bomb blew up in the VC's faces, they'd think twice the next time. If nothing else, they'd have to check every smoke grenade that they captured.

In the late afternoon we were picked up and flown back to Lam Son. General Thuan had risked having us out one night, but it was still too early for him to make it two nights in a row.

The next morning when I walked into Major Clark's office, he invited me to have a seat. I had just taken a drink from my coffee cup when Clark dropped his own bomb on me. Captain Hung was being transferred and given the command of a battalion. As the hot coffee burned through the front of my camouflaged fatigues, I sputtered that the general couldn't do that. It didn't make any sense to give the division's best unit commander an infantry battalion, not when Recon Company's weapons and aggressiveness made it more powerful than any battalion. We killed four or five times as many Viet Cong as any of the larger battalions.

Without interrupting me, Major Clark listened as I paced back and forth, giving every argument I could think of. When I ran out of breath, he quietly pointed out something I had overlooked. Captain Hung had been commanding the Recon for a long time and had been repeatedly wounded—several times seriously. At the rate he was going, he'd probably be dead within a year. He hadn't asked for the transfer, but General Thuan had decided that he deserved a break. It was hard for me to argue with that. I didn't want to see Hung killed, but I sure didn't want some political appointee taking over my company.

I left Clark's office and walked across the quadrangle to find Colonel Sonstelie. He was in his office and motioned for me to come in. "I suppose you've just gotten the word on Captain Hung's transfer. What do you think?"

"I don't like it one damn bit, but I'm glad that Hung will have a better chance of getting out of this business alive. Obviously I don't have any control over it, but by God I want a say in who replaces him. I won't work with a damn incompetent or a politician."

"I think you'll like Hung's replacement. General Thuan says he's a sharp young officer and is confident he'll do a good job."

When I got to the Recon compound, Hung was already packing. Because of all the Tet fighting, we weren't even going to be able to give him a going away party. I walked him out to a jeep and we shook hands. There were tears in both our eyes when he climbed into the jeep and drove out of the compound.

I walked back to the division area and saw General Thuan talking to a young Vietnamese lieutenant. Thuan spotted me and beckoned me over, using the Vietnamese gesture that is similar to the way small children wave bye-bye. He returned my salute, shook my hand, and introduced me to Second Lieutenant Nguyen Chi Hien, the new 5th Recon commander.

Jesus, was I unimpressed! The kid was about 5'2" tall, wore gold metal-rimmed glasses, and looked to be about sixteen years old. I was shocked when I found out that he was actually twenty-eight, a year older than me. He looked every inch the school teacher I later learned he wanted to be when the war was over. *What the hell was happening*?! Thuan had taken away an experienced, tough, and hard-charging combat commander and had given me a baby-faced schoolteacher!

I introduced myself, using a formal Vietnamese greeting, but I quickly excused myself saying that I had to meet with Colonel Sonstelie. As soon as I was inside Sonstelie's office, I blew my top. I told the colonel that there was no way I was going to nursemaid a snot-nosed kid while a 120-man recon company fought a 300-man VC battalion.

Sonstelie put up with my ranting and raving with much more patience than I would have shown a first lieutenant if I had been a full colonel. When I finally stopped, he asked me if I'd like to come in out of the field. "You've been out there almost ten months, and your unit is always getting into heavy contact. Don't you think it's about time you came in?"

I wasn't ready for that, so I changed my tune and said, "Naw, I'll give this kid a try—but if he fucks up, I'll shoot him myself and Thuan can find me another one."

I figured that Sonstelie owed me one, so I hit him with a request. One of the young American soldiers assigned to guard Gosney Compound had asked me for a job. He was tired of sitting in a perimeter bunker every night and staring into the darkness. That was his only function, and he didn't want to go home after a year in Vietnam and admit he'd only been a REMF. He was a sharp-looking kid, but our policy was to accept only people who had at least six months in combat. When I mentioned it to the others, they immediately vetoed the idea, but the more I had thought about it, the less objectionable it became. If I got this kid, he could be my radio operator. Then I —and whoever went with me (Glen or Chuck)—wouldn't have to take turns carrying the radio. We'd both be free of the damned thing, and the kid could follow me around with it.

I told Sonstelie about the young guard and said that I wanted him to be my radio operator. "It's a real pain in the ass to have to fight the VC and hump a damn radio too. Now that I've got to worry about a new counterpart, I could use the freedom the kid would give me."

To my surprise, the colonel said yes, and I walked out of his office having won a small victory. I went back to the team and told them that the Recon had a new commander and that we had a new radio operator. After pissing and moaning about losing Hung, they began complaining about taking on a kid with absolutely no combat experience. Glaring at them, I said that if they didn't agree, I would tell Sonstelie that I had changed my mind. "But from now on, one of you will hump the radio. I am never going to put that heavy fucker on my back again."

The argument stopped immediately, and I walked to the hootch where the perimeter guards lived. I found young Sergeant Collins and handed him an M-16 rifle. "Get rid of that M-14, get down to the village, buy a beret and some camouflaged fatigues, and move your stuff in with Septer and Zollenger. Have them help you fix up some decent web gear and be ready to move out at first light tomorrow morning. You're my new radio operator."

The next day we were off again, this time air-assaulting into an

LZ a half-dozen kilometers northwest of Phu Cuong near the Saigon River. We were supposed to move east from the LZ toward three blocking positions that had been established by the 1st Cav Squadron, the 1/8 Infantry Battalion, and the 3/9 Infantry Battalion. This was another plan that looked good on paper, but because we landed at high tide, the whole damned area was flooded. It took us three hours to move the first one-and-a-half kilometers. At that rate there was no way we were going to surprise any VC.

We finally hit higher ground and were making much better time when one of our Bird Dogs reported that a body was lying in a small open area not far from where we were. To make sure that the guy was not just pretending to be dead, the pilot put a WP rocket near him. He didn't move, and the rocket gave us a direction and distance to the clearing.

Just in case it might be some sort of trap, we held at the edge of the clearing and watched the area for a few minutes. No one had binoculars, but my rifle scope served the purpose just as well. After scanning the far tree line, I aimed the scope at the body. There was something not quite right about it. Instead of a normal rounded or bloated body, it looked more like a thick field jacket which was draped over a small mound of earth and molded to the contours of the ground. Hien sent two squads along the edge of the trees to ensure that we weren't being set up for an ambush, and when they met on the opposite side of the clearing, he and I walked over to the body.

It was the damnedest body I had ever seen. He had been killed by some sort of explosion, possibly artillery or a booby trap, which had blasted his rib cage and broken most of his bones. This was why he was as flat as a pancake. Instead of rotting, the sun had dried his flesh to dark brown leather. One of the soldiers turned him over with his boot. The side of the body that had been next to the ground had been completely eaten away by maggots, many of which were still crawling in and out of him.

When the soldier turned him over, the body looked and sounded like a large turtle shell and sent up a foul odor. We all stepped back and decided to leave him and get the hell out of there. I radioed the pilot and told him what we had found. "He's been dead at least a couple of weeks and looks to be about eight or nine years old."

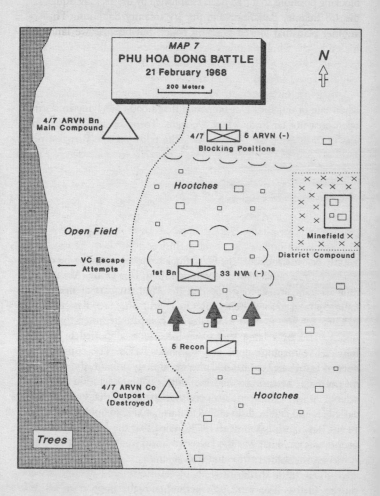

MAP 7
PHU HOA DONG BATTLE
21 February 1968
200 Meters

N

4/7 ARVN Bn
Main Compound

4/7   5 ARVN (-)
Blocking Positions

Hootches

Minefield

District Compound

Open Field

VC Escape
Attempts

1st Bn   33 NVA (-)

5 Recon

4/7 ARVN Co
Outpost
(Destroyed)

Hootches

Trees

16 A PURPLE HEART

Shortly before midnight, the 4/7 Battalion started receiving incoming mortar rounds on its main compound in Phu Hoa Dong. At the same time its company-sized outpost in the southern part of the village radioed that it too was being mortared and taking recoilless rifle and RPG rocket fire. A short time later the company reported that it was under attack by a large Viet Cong force, one supported by at least four or five machine guns and numerous RPGs. The company requested assistance, but beyond offering artillery support, the 4/7 could do nothing; the mortars had the battalion pinned inside the compound. At 0045 the company commander radioed that he had many casualties and that he didn't think that he could hold out much longer. In his last transmission at 0110, he cried that the VC were coming over the berm. That was the last word heard from the company, and it was assumed the compound had fallen.

At 0730 the next morning, about 150 Recon and Recondos were on the airstrip. We were only getting enough choppers to lift 100 men, but Hien and I wanted the other fifty to be standing by in case we needed them later. We had the aviation lift company for the whole day, so we could bring the rest our force in whenever it might be required.

None of us were very happy with our mission. Why couldn't the 4/7 Battalion go check out what was left of the compound and police up the bodies? It was an "ash and trash" job. Although we might run into a few snipers and stay-behinds, it was very unlikely the VC battalion commander would hang around to do battle. He had knocked off an ARVN rifle company, and that was a pretty good victory. The 4/7 should handle the mundane work while the Recon hunted down Charlie and exacted revenge.

When the choppers arrived, I coordinated with the aviation commander and told him to insert us right outside the 4/7's main compound. We could land anywhere in the large open area along the west side of the village, but the safest place would be next to the compound. If we tried to come in near the overrun outpost, we might take some fire from VC stay-behinds. It would only take one RPG rocket round to screw up a routine operation.

At 0825 the choppers set down in the field next to the 4/7's front gate. While the platoon leaders assembled their troops, Hien and I walked over to talk with the battalion commander and his advisors. They were pretty upset about the lost company outpost and the casualties they had taken. One of the company's platoons had been out on night ambush, so it hadn't been involved in the battle. Some soldiers in the outpost had been able to escape when the VC came over the berm, and wounded soldiers were still straggling into the main compound. They didn't have exact figures yet, but they estimated that company headquarters and its two platoons had lost about thirty-three killed and wounded.

The main compound sat about midway up the western edge of the village, about 100 meters out into an open field. Approximately one kilometer to the southeast was the Phu Hoa district compound. It too was at the edge of the field and was surrounded by a minefield and barbed wire. The overrun outpost was another 400 meters southwest of the district compound.

I wanted to see how Lieutenant Hien was going to handle this, so I kept quiet while he developed his plan. Since most commanders probably would just move directly to the objective, I was surprised when he showed the ARVN battalion commander what he was going to do. He'd first take the Recon east through the village and then swing around to the south and finally come up on the outpost from the south. This meant a circuitous walk of almost three kilometers, but any VC that might have stayed behind probably would be surprised when we showed up in their rear. They would have seen us land at the main compound. When no one showed up after an hour or two, they probably would think that we had gone off looking for the Viet Cong main body—which is exactly what we should have been doing. Hien asked the battalion commander to set up a blocking position just south and east of the main compound and to hold until we had searched the outpost. If there were any VC around the outpost, we'd have them pinned between the Recon, the district compound, and 4/7's block. The only way left for them to escape would be to cross the 300 meters of wide-open field. Both the Recon and the 4/7 could cover that with fire. It was a good plan and I complimented Hien on it.

As we folded our maps and got ready to move out, Lieutenant Dave Hendricks asked if he could go with us. Hendricks was the 4/7's assistant battalion advisor and seemed like a good guy. I told him that I thought we were going to waste our time, but if he wanted to tag along, he was welcome. It was midmorning by the time we had covered most of the distance to the outpost, and Hien decided to stop for a little while to give the troops a chance to eat. We had about another 300 meters to go, and if we were going to have any sort of fight, we wanted to have it on full stomachs.

I had just finished eating my cold rice when three Recon soldiers came around the corner of a nearby cinderblock house. One of the soldiers was wounded, and when the other two laid him down, I saw blood coming from his buttock. He said that he had been standing next to a hootch when somebody shot him in the ass. He hadn't seen where the round came from, but one of his buddies said that he thought he knew. I turned to Hien and said, "It looks like the VC left a few people behind to harass us."

I told him that I'd check it out and be back in a few minutes. He started to protest that it wasn't my job, but I just patted my rifle scope and smiled. I followed the soldier for about 100 meters through the hootches, until he stopped behind another house. We peeked around the corner and he pointed to a fairly large, shedlike building. It was a small tile factory and had stacks of red terra cotta tile all along the wall facing us. Whoever was in there was well-protected, and I might as well throw rocks as try to shoot at him with an M-16. I pulled back behind the house and looked around for a soldier with a LAW.

After I found someone, I took a deep breath and calmly walked out from behind the hootch as if I was going to take a piss. I figured the Charlie would probably wait for more people to appear before he opened up. When I was clear of some tree branches, I stopped, turned, and quickly put the LAW on my shoulder. Taking fast aim, I squeezed the lever and fired the rocket into the factory building.

When the smoke and dust cleared, I saw a large jagged hole near the peak of the tile roof. I had hit the building about ten feet above where I wanted to, and I was a little embarrassed when I heard whistles and catcalls from the troops behind me. I may have missed, but the rocket blast should have stunned whoever was in the factory, so with a half-dozen soldiers, I charged the building.

When we got there, we didn't find one VC sniper—we found *six dead NVA soldiers, an RPD machine gun, and five AK-47s!* They were huddled behind a four-foot thick stack of terra cotta tiles. I probably wouldn't have gotten them if my rocket hit where I had wanted it to go, but when it exploded over their heads, it perforated them with shrapnel and pieces of roofing tile. The troops were amazed, but I pretended the rocket had gone exactly where I had wanted it to go.

We searched the bodies and found documents that showed that these people were members of the 1st Battalion, 33d NVA Regiment. Although I was happy about killing six bad guys, I was also a little pissed. By insisting that we go into Phu Hoa Dong, General Thuan had caused us to miss an opportunity to do battle with a tough NVA unit. The 33d NVA wasn't a ragtag local-force unit. It was a well-armed and well-trained regiment.

A couple of hundred meters farther on, we were able to see what was left of the outpost. It had obviously taken a lot of mortar, rocket, and recoilless rifle fire, and there were several dead ARVN on the edge of the berm. We hadn't run into any more NVA, so it was possible that the six whom we had killed had been withdrawing when we bumped into them. Hien sent a squad of soldiers to check the outpost, and they returned a few minutes later saying they had found only ARVN bodies.

We still had to move 600–700 meters to reach the edge of the district compound's minefield, and we probably had to cover another 200 meters to link up with the 4/7. If we didn't run into any more stay-behinds, we should have this thing wrapped up in an hour or two.

For the past hour and a half, we hadn't seen a single civilian, and this part of the village was like a ghost town. We began searching the hootches and found no one in the family bunkers. This was not particularly unusual in an area that had just had a big fight, but it was still a little unsettling. Normally, the peasants return to their homes when they believe the fighting has ended and it is safe to do so. Obviously, these villagers still considered this area to be dangerous, and that suggested there might be some fighting yet to do.

We had barely moved 100 meters beyond the outpost when bullets began cracking between the empty hootches. It wasn't a massive amount of fire, but clearly there were at least a couple of squads of NVA stay-behinds between us and the 4/7's blocking position. The troops took cover behind the hootches and began returning fire, but it was almost impossible to pinpoint the NVA riflemen. They were well camouflaged and could be in any of the hootches or bushes to our front. I got on the radio and called Division, asking for some gunships and one of Hooker's Bird Dogs. Maybe the aircraft could help us spot where the fire was coming from, and the gunships could give us a hand in knocking out the NVA snipers.

Because the gunships were standing by at the Lam Son helipad, it was just a matter of a few minutes before I had both Hooker and the fire-team leader calling me on the radio. I tossed out a smoke grenade to mark our positions and briefed the two pilots on our situation. Hooker made a couple of low passes, but the NVA were too smart to give themselves away by firing.

I called the gunship leader and asked him if he had a good fix on everybody's positions. He had the 4/7's location, but he wanted a better mark on the Recon's forward trace, so I had Hien pass the word to pop smoke at either end of our line. The first rockets were too far to the north, so I told the gunships to move their next pass closer to our positions. The flight leader was a little concerned, but I assured him that we had plenty of cover behind the hootches and trees. Nevertheless, the gunships continued to put their rockets too far from us to do any real good. Wherever the NVA snipers were, they were close—probably no more than fifty or seventy-five meters away.

I radioed the flight leader my initials and told him that I accepted full responsibility, but just when he and his buddy were getting close, he called and said that they had to go back to rearm and refuel. I wasn't very happy with this news, not even when he said that they'd be back in less than an hour. When I told Hien that they were leaving for a while, he wasn't very pleased either.

While the gunships were gone, there wasn't much we could do. We couldn't flank the NVA on the left because of the large open area, and we couldn't get around to the right because of the district compound's minefield. It was a little frustrating, but at least we had the NVA squads cornered and they had no way to escape. It was still early in the day, so when the gunships got back, we'd go in and clean out the bastards.

There continued to be sporadic firing by both the Recon soldiers and the NVA. The Recon would periodically pop an M-79 round into a suspected NVA location, and a few AK-47 rounds would crack back from somewhere. I crawled to the edge of a building and slowly scanned the scope back and forth across the part of the NVA-held area that I could see. A couple of Recon soldiers saw what I was doing and one of them took off his helmet. Putting it on the end of a stick, he stuck it around the corner of the hootch that he was behind. About thirty meters in front of me, I saw a small cloud of dust rising from the ground as an NVA fired at the helmet. The bastard was apparently in a well-camouflaged spider hole. I wouldn't have located him if I hadn't been looking at just the right spot when his muzzle blast kicked up the dirt. I waved to the soldier with the helmet, and when I was ready, he poked it out again. Crack! Bang! I fired just a fraction of a second after the NVA soldier did, and although I couldn't be certain, I thought I gotten him. Just to be certain, I crawled back around the hootch and borrowed an M-79. I put two explosive rounds in the bushes and figured that they finished him off. My assistant waved the helmet and pointed to a hole in it. That NVA may have been a pretty good shot, but he wouldn't have scored very high on an IQ test.

An hour passed, and still the gunships hadn't returned. I called Division to find out what had happened to them. The people there didn't know, but they said that they would check it out and get back to me. They asked about our situation, and I told them that we were waiting for the choppers to give us a hand in cleaning out what I estimated was about ten or fifteen NVA.

Twenty minutes later I called again, but they still didn't have a status on the gunships. I told them that we were sitting there with our thumbs up our asses while the damned chopper pilots were probably having lunch. I demanded that they find out when the birds were coming back. It was still early, but we didn't have all day.

Everybody was getting antsy, so after another fifteen minutes, Hien got up and said that we'd attack without the gunships. He radioed the platoon leaders to pass the word. When they called back, he gave the command to assault. With a massive burst of automatic fire, everyone charged forward into the hootches and brush in front of us.

We'd only gone a few meters when the air was filled with bullets and explosions. *There were fucking NVA everywhere*! What we had thought was probably only a couple of squads turned out to be at least a reinforced company! The bastards were under our feet, in the hootches around us, and to our front. I had been in some close-quarters fighting before, but this was downright hand-to-hand! We were killing NVA all around us, while they were shooting back with AK-47s, machine guns, RPGs, hand grenades, and everything else they had. Within a few seconds I heard a 12.7mm heavy machine gun open up, and I knew that we had really stepped in some deep shit!

If the NVA hadn't been dug in over a wide area and had it not been for all the cinderblock houses, we wouldn't have lasted two minutes. The NVA couldn't bring all their fire on us at the same time, and the well-built houses provided us good cover. It was obvious that we were heavily outnumbered and weren't going to be able to win this battle by ourselves. Hien passed the word and we began to fight our way back the way we had come, dragging our dead and wounded with us.

We reestablished a line at the same spot we had started the attack. We were still taking some fire, but were able to keep the NVA from reoccupying their old positions immediately in front of us. I got on the radio and called Division. "It looks like we've got more than we thought. We have at least one reinforced November Victor Alfa [NVA] company and probably two. They've got at least one twelve-point-seven heavy machine gun and possibly recoilless rifles. We need an airstrike as soon as possible and will need at least one Dust Off as soon as we can collect our casualties."

I gave the handset back to Collins and looked around for Glen. He was behind a house a short distance away, so I yelled for him to get over to me. When he crawled up, I told him that I wanted him to go back and find a place where we could bring in a medevac chopper. Checking with Hien, I told Glen to take one of the Recon radios so he could talk to both the Dust Off and me.

He didn't like leaving the fight, but we really didn't have any alternatives. I couldn't send Collins because he'd never done this before, and the Dust Off wouldn't come in unless the pilot was talking to an American.

"What about Lieutenant Hendricks? Can't you use him?"

I'd completely forgotten about Dave in the heat of the battle, but when I looked around, I spotted him a few meters away. I gave him a thumbs-up and he waved back. Turning back to Glen, I said, "Hendricks is a guest. The Recon is our company, and we're supposed to take care of it."

Swearing a blue streak, Glen grabbed a radio and stormed off with one of the Recon soldiers. He snarled over his shoulder that when he found a good spot, he would send the soldier back to let us know where it was.

The air force FAC showed up much quicker than I had expected, and I radioed our situation. I also warned him that Charlie had a 12.7mm heavy machine gun, so he better keep as much altitude between him and the ground as he could. When he asked us to mark our position, I turned to Hien and told him to have the troops pop smoke. When the FAC said that he had a line of two yellows and one purple, I confirmed that those were ours. After he got a fix on the 4/7's position, I asked him what ordnance his birds were carrying. In a minute or so, he radioed back that they were carrying 500-pound GPs and napalm.

"Roger, understand. My counterpart will not withdraw any farther Sierra [south] than is absolutely necessary, so you're going to have to be very accurate. I want you to put the iron bombs in first to blow down some of those hootches and brush, and then lay the napalm right on the ground."

I told Hien that we would have to pull back at least another 100 meters. Even then we were going to take a lot of bomb fragments and heat from the napalm. Everybody would have to get behind some good cover to keep from becoming casualties from our own airstrike.

Reluctantly Hien agreed, but he said that he wanted to get into the NVA's positions as soon as possible after the last bomb was dropped. I was all for that and asked the FAC to have his Phantoms make a couple of 20mm strafing passes so the NVA would go to ground while we pulled back. I didn't want Charlie to hug us when we pulled back. I wanted him to stay in the target area.

Hien and I were lying behind a cinderblock hootch that gave us good cover, but every time a bomb exploded, it bounced me several inches in the air and I came back down on my ammunition pouches, bruising my waist and chest. Glen called on the radio and said that he was bringing in a Dust Off under cover of the airstrike. The casualties were mounting and it would take several flights to medevac them all.

"Roger. It looks like you're going to be busy, because we're going back into them when the airstrike is finished."

A few minutes later the FAC radioed that each of the two Phantoms would make one more napalm run, and then they would be done. I told Hien, and he radioed for the platoon leaders to be ready to attack after the last pass.

The heat from the napalm could be felt even behind the hootches, and it reminded me of the experience of opening an oven door to see if the turkey was done. I hoped that it was cooking Charlie's goose, but I had been through this too many times before to expect that we had finished him off.

As soon as the red flame disappeared and the black napalm cloud billowed up, Hien waved his hand and we started back toward the NVA position. Everyone was firing automatic and making so much noise that it was hard to tell if we were getting any return fire.

I don't know why bombs and artillery rounds make such impressive explosions—yet never seem to kill all the bad guys. It must be one of Murphy's Laws that short rounds always hurt a lot of friendlies but that bombs and rounds that hit on target only piss off the VC. The bastards recovered quickly and were still full of fight.

Like everyone else, I had my M-16 on "rock and roll" and was firing from the hip. This was one of the few times I had actually seen the VC up close, and they kept popping up all over the place. The NVA battalion had spent half the night digging spider holes and preparing bunkers. Everywhere I turned, there was an NVA soldier or a dust cloud from a muzzle blast. The air was filled with bullets going in all directions and punctuated with explosions from grenades, M-79s, and RPG rockets.

A burst of machine gun fire cracked by me and I hit the dirt. Luckily I found myself lying in a very shallow depression that gave me the absolute minimum of cover from the gun firing at me. Had I been a couple of feet away, I wouldn't have had any cover at all. I could hear the gun firing, and it was only ten or fifteen meters away. I turned my head to the rear and saw that while I had been so absorbed in acting like John Wayne, the Recon had pulled back about 100 meters. I was all alone and lying closer to the NVA than to my own people.

If I could get to one of my grenades, I might be able to hurl it into the bushes where the machine gun was hidden. Both grenades were attached to my ammo pouches, and they were under my belly. As I was fumbling with one of them, the machine gunner spotted the movement and blasted the weeds next to me. I pressed my face into the ground and felt the bits of grass and dirt blowing across my back. I was completely pinned down and couldn't move at all.

After a few moments the gun stopped firing and I lay there staring into the dirt. I wasn't scared, but I was thoroughly pissed at myself. Only one thought went through my mind—"Parrish, you really are a dumb shit. It's just a matter of time before the VC tosses a grenade on top of you, or the gunner picks his machine gun up just a little —and it's all over."

I had no options. I couldn't run, crawl, shoot, or throw a grenade. I couldn't do anything, and the VC weren't going to wait much longer before they finished me off. If I moved one more time, they'd know that I was still alive—and "that would be all she wrote."

I was still silently cursing my stupidity when there was a big explosion and I felt something hit me in the neck, just below the base of my skull. Suddenly, everything was crystal clear—*I had to move now*! Whatever the explosion was, it must have sent shrapnel into the machine gun position. If the crew hadn't been wounded, at least they would have ducked.

I leaped to my feet and saw an NVA soldier kneeling about fifty meters away, holding an RPG-7 on his shoulder. He was the bastard who had tried to blow me away. I could have kissed him, but instead I blasted him with my M-16. Then I sprayed the bushes where the machine gun was hidden. Turning, I ran as fast as I could toward a hootch behind me. While I ran, I held the M-16 over my left shoulder and sprayed the rest of the magazine in the general direction of the NVA.

When I cleared the corner of the house, I was completely out of breath. I leaned against the concrete wall and gasped for air while NVA bullets chipped away at the edge of the house. That damned machine gun was still trying to kill me. As soon as I caught my breath, I jerked the LAW off my shoulder and blew the machine gun sky-high—it was the most satisfying rocket that I had ever fired.

Now that I was safe for the moment, I put my hand up to my neck. When I pulled it back, it was covered with blood. My neck was numb and there was a hell of a ringing noise in my ears, but I felt no pain. I turned around and saw my troops coming toward me—they were attacking again. I put a full magazine in my rifle and chambered a round. Then when the soldiers were almost to me, I charged around the side of the house, and back toward the NVA.

It was the same thing again; the enemy troops were putting up a hell of a fight, but we were pushing into them. This time we got about a hundred meters farther along before our attack bogged down again. There were just too many of them and they were too well armed. On top of that, we were running low on ammunition. I heard a whistle and saw our soldiers slowly begin withdrawing.

We pulled back to just about where I had been wounded, and it took me a few minutes to find Hien, Collins, and the command group. When I got to them, they all just stared at me. Then Collins said, "Shit, sir! You've got blood all over your neck and shoulder!"

As Collins tied a battle dressing around my neck, he told me that he had watched me haul ass after I'd been hit. "The machine gun bullets were hitting the ground right behind you, all the way back to the hootch."

"Well, I hope you paid attention, because only a dumb shit gets himself into that kind of situation. I wasn't paying attention, and I'm damn lucky I'm not going back to Lam Son in a body bag."

When Collins finished with the bandage, Hien started on my case. "I told you before we had soldiers to fight. If you get killed, how can we get support?"

He went on to say that they hadn't been attacking the last time. When they saw that I was all alone out front, they went charging up to rescue me. But just as they reached me, I took off toward the NVA. I was a little stunned and replied that I hadn't realized that. I thought they were attacking, so I just went along. Just then Glen called on the radio and said that the last Dust Off had been hit by 12.7mm fire and had crashed. The crew was okay, but the chopper was disabled.

"Great! Just when I need a Dust Off, you tell me we ain't got any."

"Have you been hit?"

"Roger, but I'm okay. I'm not leaving until this business is all cleaned up."

Hooker was still circling overhead and had monitored my radio message to Glen. He asked me how badly I'd been wounded, and I told him the same thing I had told Glen. I personally didn't need a Dust Off, but I had people who did. I also needed another airstrike; even more vital was a resupply of ammunition, although I didn't know how Division could get it to us.

As soon as I released the push-to-talk switch, I heard Division calling. They were talking to Hooker, and they wanted to know how badly I was wounded, what my situation was, how many casualties we had taken, and half-a-dozen other things. (My radio battery was getting low, so the division TOC couldn't hear my transmissions very clearly and were using Hooker to relay the information.)

Hooker told them what he knew and said that we urgently needed another airstrike and some ammunition. They acknowledged the request and said that they would get right back to him. In two or three minutes they called him again and told him that I was to evacuate myself immediately.

Christ! My company's got all kinds of casualties, we're mixing it up with most of an NVA battalion, we're running low on ammunition—and they want me to evacuate myself?! Somebody back in Lam Son has gotten his priorities all screwed up. Hooker called me and asked if I had monitored Division's transmission. Of course I had, but I said that I was having trouble with my radio—probably a weak battery or a bad handset. Hooker knew exactly what I was doing, and he told Division that he was having a hard time understanding me because something must be wrong with my radio.

Hooker radioed back to me and asked if we could hold out for a while, at least until the airstrike got there. I reminded him that we were the ones attacking the NVA—not the other way around. It wasn't a matter of holding out, it was a question of whether we could charge the enemy again before it got dark. He laughed and said that he'd be back as soon as he could. He'd try to figure out some way to get us the ammunition that we needed.

The firing had died down to just sporadic shooting, and it was clear the NVA were having a few ammunition problems of their own. As we waited, I reloaded my empty M-16 magazines. I was still okay, but I had fired more than two-thirds of the ammo that I had brought with me—and I typically carried more ammunition than the average Recon soldier.

About thirty minutes later, Hooker called and said that he was airborne again and on his way with as much ammo as he could fit in the back of his Bird Dog. I told Hien that I was going to get some ammo and, together with Collins, moved about a hundred meters to the rear. When Hooker was ready, I popped smoke. He came in just above the treetops and kicked out a case of M-16 rounds. It crashed through the low trees and smashed into the ground only a couple of meters from where I was standing. I told him that he had a good spot, but I asked him to kick the next ones out a little farther to the north—I didn't want to be killed by a falling ammo box.

On his second pass the NVA began shooting at him, and I could hear the Recon trying to suppress them. Despite the relatively heavy VC fire, Hooker made several more passes, until he had only one case of M-79 rounds left. When he told me what he was about to drop, I quickly told him to take that one back to Lam Son. I wasn't going to risk having the rounds explode when they hit the ground.

When Hooker was finished dropping everything but the M-79 rounds, I moved back to the command group and told them to distribute the ammunition to the troops we had left. I didn't know how many casualties we had taken, but it was clear that it had been quite a few. Then Glen called on the radio. "I don't know how many of our people have been hit, but the number is mounting up. When you bring in the next airstrike, I've got a couple of choppers standing by to come in and get some of them out. They'll be bringing in some Recondo reinforcements and additional ammo."

As soon as Glen finished, Division was on the air—this time *ordering* me to evacuate myself on the next chopper. I didn't even bother to acknowledge the transmission, and I told Collins not to answer any more calls.

The Recon's forward observer was on the ball, and when Hooker was clear of the area, he began putting artillery into the NVA positions. The artillery fired for about fifteen minutes, and then the FAC called. He was en route with another pair of F-4s. It was the same guy as before, so I didn't have to spend much time briefing him. We popped smoke to mark our forward trace, and he lined up his Phantoms for their east-west passes. As the birds began dropping their ordnance, our platoon on the left reported that small groups of NVA were breaking and running across the large open field. Apparently some of the VC had had enough and were trying to escape.

I told Collins to follow me and headed to the edge of the village, ducking behind hootches whenever an F-4 made a bomb pass. When

231

I got there, I found a machine gun crew and several soldiers. Originally the field had been covered with large trees, but in July, Rome plows (large bulldozers) had cleared it. Now it was bare except for some large uprooted tree stumps scattered here and there. The gunner explained that the NVA were trying to make it across the field by running from one stump to another. There were four large piles of dirt and roots that seemed to be the most popular. One or two NVA would break from the edge of the village and race about fifty meters to the first stump. After catching their breath, they'd haul ass another fifty meters to the second one. The last two stumps were about a hundred meters apart.

As he finished describing the situation, I spotted two khaki-clad figures breaking from behind a hootch and charging across the field to the first stump. I asked the gunner why he hadn't fired, and he replied that he couldn't see the first part of the run and had to leave it for the riflemen.

This was too good to be true. You rarely saw live VC, and to have them running across a 300-meter-wide field in broad daylight was really something. I told the riflemen that I'd take over for a while and carefully laid my M-16 on the pile of dirt in front of me. I had just gotten ready when another Charlie broke for the stump. He was about 200 meters away, and I put the scope's cross hairs just a little in front of his chest. I squeezed the trigger and the rifle kicked. When I brought the scope back on target, he was nowhere to be seen, but the cheering of the soldiers told me that it was a good kill.

For the next ten or fifteen minutes, the machine gun and I took turns firing at the NVA. If more than one NVA soldier made the first run or if I missed, the machine gun picked them up when they tried to get the rest of the way across the field. There were so many VC trying to escape in that short period of time that I knew we had the 1st Battalion, 33d NVA Regiment by the balls. Either their commander had ordered them to try to break out or they were panicking. Whatever the reason, we had them in a vise. They were cut off on the north side by the 4/7 Battalion, and they were pinned in by the district compound and its minefield on the east and by the open field to the west. Between the 5th Recon, the airstrikes, and the artillery —Charlie was hurting.

The FAC called and said that his aircraft were about to make their last passes, so I reluctantly turned over my sniping duties to the others and made my way back to the command group. Because the F-4s were now dropping napalm, Hien knew that the airstrike was almost over, and so he ordered his people to prepare to move. I wasn't looking forward to another assault. My neck was beginning to hurt like hell, and I had a splitting headache. More important, the illusion that I couldn't be killed was gone, and the reality of death had replaced it. I didn't pull any John Wayne stuff this time—before I moved from behind one obstacle, I made certain that I had another covered position that I could get to quickly.

We advanced another seventy-five meters into the NVA's position, but even after three assaults, two airstrikes, and a half-hour of artillery, the bastards still outnumbered us—and they were fighting for their very lives. Most of their holes and bunkers had overhead cover. The bombs did a lot of damage, but unless the artillery got direct hits, it only kept the NVA in their holes; it didn't kill them.

There was a lot less NVA fire, but it was still more than we could overcome. We were pretty much out of hand grenades, M-79 rounds, and LAWs, and we were reduced to only M-16 and M-60 bullets. These weren't much good against a dug-in enemy, so Hien ordered another withdrawal and we slowly pulled back. We had to get the right weapons to go after the bunkers—rifles and machine guns weren't cutting it. As soon as we broke contact, the firing died away. Neither we nor the NVA had any extra ammo to waste.

The artillery started up again, and I called Glen to find out the status of our resupply and reinforcements. He'd gotten all the seriously wounded evacuated and was working to get the others out. Apparently the 12.7mm heavy machine gun had been knocked out, because none of the choppers had taken any fire when they came in to land.

I made my way back to where he had set up, passing Recondos who had been brought in on the choppers. Fresh and loaded with ammunition, they were a welcome sight. When I reached the LZ, Tony was in a chopper, kicking out several cases of ammunition and grenades. When he was done, some Recondos began helping wounded soldiers into the helicopter. Tony spotted me and made his way through the wounded soldiers. "How badly are you hurt?"

233

"I'm okay. Where are the rest of the Recondos?"

"Division released the gunships and the lift company just before noon, so we only got a couple of admin choppers to move the Recondos and the ammo here and take your casualties back."

I started cursing, but before I could say much, Tony interrupted me. "I've been told to personally give you a message. If you don't get on the next bird back to Lam Son, you're going to be reassigned to Division G-5 [Civil Affairs]."

"Sonstelie wouldn't do that."

"Yes he would. He's worried about you and is thoroughly pissed that you won't answer their calls. He wants me to make certain you understand that you'll never go to the field again—he'll have you in charge of building schools and digging wells. I think he's serious, so you had better get your ass on the next chopper."

"Bullshit! After this is over, we'll just tell him you couldn't find me until it was too dark to bring any more choppers in."

"He's already figured out you'd say something like that. He said that if you're not back by 1800, you're going to be the civic action chief even if I hadn't told you in time. You'd better get on the next bird."

"How the hell is that going to look?! My Vietnamese are fighting most or all of an NVA battalion—and I go and evacuate myself!"

"Don't worry about it. First, Division is passing the same message to Lieutenant Hien, so he knows you're not leaving of your own free will. Second, they're ordering the Recon to pull back and set up a block for the night. General Thuan doesn't want them to attack again until first-light tomorrow. By then the VC will have disappeared into the Triangle. So there's really no reason for you to stay—except for your fucking pride."

"All right, but don't let the Recon do anything until I get back. I ain't hurt bad and I'll be on the first chopper out here tomorrow morning."

I felt like hell when the helicopter lifted off. Although all the seriously wounded had been evacuated, there were still some lightly wounded soldiers who had to wait for the next bird. I felt like a captain who hadn't been allowed to be the last to leave his sinking ship.

When the chopper landed at Lam Son, two guys with a jeep were waiting for me. Tony had radioed that I was on my way, and they were there to take me to the dispensary. I told them to forget the medics and take me to the TOC instead.

As soon as I walked in the door, I started raising hell. "Who was the motherfucker who released my gunships and the helicopter company?! When I needed them to bring in my Recondos and an ammo resupply, they were gone!"

I guess that with the bandage and all the blood and dirt, I looked pretty wild, because everybody just backed away from me. (Vietnam was a place where you learned to be very cautious of an armed and angry grunt.)

Colonel Sonstelie waited until I had blown my head of steam and then walked over. "Damn Bob, we thought you'd had it. We've been worrying like hell about you. We heard you tell Sergeant Septer that you'd been hit in the neck, and I had a medic explain what that might mean. He said there was no such thing as a *minor* neck wound and you could become paralyzed if that chunk of steel got to working around inside you."

His last statement stopped me cold. I hadn't considered that the hole in my neck might be serious. I calmed down and gave him a description of the action and of how we had attacked the NVA battalion, thinking that it was only a platoon or two.

"Do you mean you didn't know there was an NVA battalion between you and the 4th Battalion?"

"Shit no, colonel! Sometimes we do stupid things, but the Recon is not dumb enough to attack an NVA battalion, not with only a hundred men! The way the NVA set up, they expected an attack from the north, where the 4/7's compound is. Because Hien took two hours to circle around behind them, they were caught with their pants down. Their bunkers' firing ports were all aimed at the 4/7, and they had only outposted their rear. We were in the middle of them before either of us knew what was happening. By that time they were having problems reorienting their weapons. When they did, they found they were shooting into the center of their own perimeter. It was a real goat fuck, and they were as confused as we were."

"How come you didn't have the 4/7 attack in the north?"

"Hell, sir, with all those RPGs, recoilless rifles, and .51-cal machine guns, Charlie would have made short work of the 4/7. They would have done *us* a job if we hadn't hit them from the rear."

I asked him if he had any count on our casualties. One of the captains said that so far the choppers had evacuated nineteen dead and wounded, but there were more to get out. Now that I had calmed down, I was really feeling shitty. Besides the pain that was beginning to get to me, I was hurting for my people. I had left that morning with 100 good men and had lost about a quarter of them.

Two medics showed up and I climbed in their jeep for the ride to the dispensary. When I walked in the door, a real nonmilitary-type guy was waiting with his tools. I took off my shirt and laid down on the table. Before he started he said, "I'd prefer not to use an anesthetic because I'm concerned about nerve damage. Is that okay with you?"

"Yeah, Doc. Do whatever you think best."

That was a mistake, because I spent the next fifteen minutes gritting my teeth as he probed around inside my neck. I don't know why he didn't think the wound was big enough, but he took a scalpel and made it bigger. Just about the time that I was going to demand some anesthetic, I heard a clang as he dropped a chunk of shrapnel into a metal bowl. He probed around a little bit longer and then said, "I think we've got it."

After sewing up my neck and putting a proper bandage on it, he showed me a small chunk of steel. It still had some paint on it—the baby-shit brown color the Red Chinese Army uses. "You're one lucky guy. If that piece of steel had been a centimeter higher, it would have penetrated your skull and brain. If it had been a little deeper or more to the right, you'd be paralyzed from the neck down. I'm going to give you something for the pain, and tomorrow morning we'll evacuate you to Long Binh. The 93d Evac Hospital is equipped to handle head wounds, and I want them to check you out."

The next morning I still had a headache, and my neck was swollen and throbbing with pain. Somebody brought a jeep and drove me to the chopper pad, where I climbed into an admin helicopter which flew me to Long Binh. I insisted upon taking my rifle and web gear, because there was no way I was going anywhere in Vietnam without them.

A gurney was waiting next to the hospital pad when we landed, but I waved it away. I wasn't a cripple, and I could certainly walk into the triage room. A medic took my rifle and gear and said that he was going to turn them in to some sort of supply room. When I was released, I could get them back.

The doctor read the medical report that our man had written and asked me a few questions. I went through the moving finger, the light in the eyes, the rubber hammer banging my shins and elbows, and then a bunch of other esoteric crap. Then they took an X-ray and a medic checked me into a ward.

I was still tired and groggy, so I was asleep in just a few minutes. They woke me for afternoon rounds, and a couple of doctors with blood on their operating gowns banged on my joints and shined lights in my eyes. Except for asking me if I had sensation when they tickled the soles of my feet, they pretty much treated me like a piece of beef. When they started to walk away, I called them back and asked them what the story was.

"The X-rays show they got all the shrapnel out. Your reflexes are normal, and I think that beyond some relatively minor nerve damage, you'll be okay. You might have a headache now and then, but nothing too serious. We'll keep you here for a couple of days, and then you can go back to your unit. You won't have to go back to the bush for a month or so."

That was good news, and it pretty much confirmed what I had suspected. My headache had gone away, and other than the pain and swelling in my neck, I felt pretty good—good enough not to hang around this place for very long.

The first thing the next morning, I got up and put on my camouflaged fatigues. A medic told me that I had to stay in bed, but I just ignored him. I walked outside and hunted around until I found the supply hut. When I told an overweight REMF sergeant that I wanted my M-16 and web gear, he tried to give me a ration of shit. Except for the Vietnamese rank on the beret in my pocket, there wasn't anything on my fatigues, so he probably thought that I was a young sergeant. I didn't bother to tell him any different—I just told him that I'd jump over the counter and beat the hell out of him if he didn't give me my stuff. The frown on his face turned to surprise when he got my equipment and read the name "Lieutenant Parrish" on the cardboard tag attached to it.

It took me several hours to find a chopper that would take me to Phu Hoa Dong. When I finally located one, I climbed on board and fifteen minutes later landed near a cloud of yellow smoke that Tony had thrown for the bird. Everyone was surprised to see me, and I warned them not to use my call sign or to make any reference to my being back with the Recon. Division was to believe I was still laid up in the hospital.

Tony and Glen brought me up to date on what had happened while I had been screwing around in the rear. We had lost twenty-three men killed and wounded—about a quarter of all the troops we had started with. The surviving NVA had been able to withdraw across the field after it got dark, and they had gotten away. The next morning, the Recon had searched the battlefield and policed up one 60mm mortar, seven RPGs, three machine guns, a BAR, and a pile of AK-47 rifles. There was a lot of confusion over the number of NVA dead, but Tony and Glen had been able personally to count sixty-four bodies. The Vietnamese had a much higher count, but my team hadn't been able to confirm it. (Hien said that his people and the 4/7 had found over 100 dead.) There were more dead than weapons, and as usual, our people found a lot of blood pools without any bodies. The 4/7 was finding more blood trails in the woods northwest of the village. There wasn't any doubt about it. We had kicked the shit out of the 1/33 NVA—but we had lost a lot of good men doing it.

The next day Division called Tony and asked him if I was with the Recon. As I shook my finger back and forth, Tony said that he hadn't seen me since I had been evacuated. Division came back on the radio: "The hospital has called and said he's [me] gone AWOL, and they don't know where he is."

There was a long pause, and then a voice that sounded like Colonel Sonstelie's said, "If you see him, tell him he can come home—all is forgiven."

While I'd been in the hospital, the Recon had policed up the battle-field. The NVA's weapons and equipment had been collected and airlifted to Lam Son, and the VC bodies had been hauled somewhere and buried, but there were still random chunks of meat scattered around—arms, legs, and unidentifiable human parts. This was the 4/7's village, so it could take care of the small details.

Our fight had leveled 900 square meters of Phu Hoa Dong. There were a few badly damaged cinderblock houses still standing, but almost everything else was rubble. I was amazed at the NVA fortifications. To prepare for the ARVN reaction force, the NVA had constructed bunkers, slit trenches, two-men fighting positions, and spider holes throughout the area; most with dirt and log overhead cover.

The biggest thing that saved our asses was the way the NVA organized their defenses. Expecting the 4/7 to come at them from the north, the NVA had arranged their positions in depth and oriented in that direction. We completely surprised them by showing up in their rear. We were able to get into them before they could readjust themselves. As I had told Colonel Sonstelie, we sure as hell wouldn't have attacked such a large, dug-in force had we known what we were up against. We and the 4/7 would have set up blocking positions and pounded the NVA with artillery and airstrikes. We had won a big one, but it had been a mistake. I wasn't happy about our casualties, but Lieutenant Hien had performed very well. Despite his school-teacher image, Hien had proved himself to be a tough and aggressive combat leader.

After a few days of well-deserved rest, we were getting ready for a two-day patrol. We were going to run a sweep just above Lam Son, and we seldom ran into anything big in that area, so I decided to take George with me. The monkey no longer needed a leash when I was around, so I just put her on top of my rucksack and walked out to meet Glen and Collins. I guess I looked a little like Long John Silver and his parrot, because I made quite a hit with everyone who saw me.

The Recon was in company formation when I walked into the compound, and I pretended not to notice when the troops burst out laughing. Lieutenant Hien grinned, but when he reached up to pet George, she moved over to the other side of my rucksack, staring suspiciously at him. I was afraid that she'd bite him, but she minded her manners.

When we found nothing over the next four hours, I was certain that this would be another waste of time. We stopped for a while and choked down our ARVN C-rations. I offered George some dehydrated rice, but after smelling it, she threw it on the ground and wiped her hand on my pants.

We finished eating, pulled on our rucksacks, and headed out again. I was just settling into a stride when a burst of AK fire cracked by the command group. I hit the ground and crawled behind a large ant hill. Everybody was down in the grass, and I could see the troops scrambling for cover. After a few seconds of silence, someone yelled that there was a sniper in the trees to our front. Hien gave orders for the 3d Platoon to try to flank the VC's position, and the soldiers pulled back and began maneuvering to the left. Again the sniper fired a burst, but this time at the moving troops.

I knew roughly where he was, but he was so well-camouflaged that I couldn't spot him. I put the scope on low power to give the widest field of view and scanned the trees. In a few seconds he fired again and I spotted some moving branches. I yelled for everyone to watch my fire and put a six round burst into the tree.

The next thing I saw were two tiny hairy hands grabbing at my sunglasses—George had panicked. She was screeching and grabbing at my glasses, beret, the bandage on my neck, and anything she could get her hands on. I kept pushing her away, but she was terrified and was climbing all over my head. I completely forgot about the VC sniper and began praying that she wouldn't piss on me.

Apparently some of the others saw where my rounds had gone, because everyone around me began firing, scaring George even more. Hien yelled something and we began rushing the sniper's tree. I tried to keep up, but it was impossible to run and pull George out of my face at the same time. When I finally got to the tree, the others were standing around a dead VC. It was just as well that they had something to occupy them, because I was a mess. It took me ten minutes to retrace my route and find my missing beret and glasses. That night Glen asked me if I were going to take her into the bush again, and I emphatically stated *no*.

We got back to Lam Son early in the afternoon, tired, dirty, and hungry. After dumping our web gear and rucksacks in the hootches, we tracked down Tony and Chuck to see if they wanted to get some fried rice with us.

A couple of hundred meters down the road from Lam Son's main gate was a place called Sherwood Forest, a combination of outdoor restaurant and whorehouse. Surrounded by a high concrete wall, it had tables under rubber trees and a great Chinese menu. There was an indoor dining room, but no one used it during the dry season. It was a fairly class operation, and the madame had a couple of tough-looking bouncers to make certain that the American GIs from Phu Loi didn't get too rambunctious.

Since George was now eligible for a CIB, we decided that we ought to take her along to the restaurant to celebrate her first—and last—firefight. The madame was a tough old bird, but she liked us because we advised her favorite ARVN unit. She wasn't particularly keen on having George in her facility, but she didn't give us any problem.

The girls at Sherwood Forest were better mannered than the prostitutes along the strip outside Phu Loi, and they didn't harass you if you were only there to eat. They'd come around to the tables, and without being too pushy, they'd ask if you'd like to buy them some Saigon Tea—real tea, but costing the same as a shot of whiskey. They were well-groomed and wore *ao-dais* and western dresses, rather than the more suggestive clothing typical of prostitutes. Instead of using names, they wore small heart-shaped pins with numbers on them. If you wanted to get to know a particular girl more intimately, you discreetly told the madame that you'd like to visit with "Number 24" or "Number 16." Of course, no one wore the number ten.

I was a normal, red-blooded young man, but there were a few ways for an officer to ruin his career: screw up in combat, cut ears off dead VC, or catch venereal disease. Nothing was said officially, but you never knew if a promotion board might look at your health records. In the past, I just smiled and said "no thanks" whenever a girl came to my table, but today things were a little different. Whenever a sweetheart came near me, George bared her teeth to warn the gal away. One girl didn't take the hint, and George bit her hand when she touched me on the arm. That took care of the whores—and it provoked some comments about my relationship with my female monkey.

George and I shared a large plate of fried rice. After we were done eating, the whole team sat around the table and sipped big glasses of beer with chunks of unpotable ice floating in it. George was really into beer, and every time I took a drink, she insisted on one too.

We were on our third beer when George suddenly fell off the arm of my chair and landed in the dirt. With four hands and a prehensile tail, monkeys can hang on anything—provided they're sober. We were all rocking back and forth and laughing, when George jumped to her feet and took off running. I leaped up and started after her, but she was making good time despite being stoned.

In seconds half the people in the place were chasing George all over the compound. Just when we thought we had her cornered, she'd take off again, going like hell. She raced into the restaurant and out through the kitchen, adding the cooks to the group of pursuers.

Behind the main building was a long shed containing the rooms where the prostitutes entertained their customers. Because of the heat, the doors to the rooms were open with privacy provided by strings of beads or hanging strips of plastic. Just when I thought we had her cornered, she disappeared into one of the rooms. In an instant we heard male shouting and female screaming, and then she came barrelling back out the door.

If I were to script a Keystone Cops movie, I wouldn't change a thing. George went down the row of doors, racing into the rooms, trying to find a way to escape. In her wake she left naked people yelling, cursing, and screaming.

After being unsuccessful in the last room, she shot across the compound and raced up a rubber tree, leaving me to face a crowd of irate people. It took a while, but I finally calmed most of them down enough so that they returned to whatever they were doing before they had been interrupted. I gave the madame a couple of thousand-piaster notes, and that seemed to mollify her enough to allow me to coax George down from the tree. As we left, she told me that my monkey was no longer welcome in her establishment.

The next morning we climbed on the backs of the tanks and armored personnel carriers. This was going to be another sweep to the south, and then we would head home again along Highway 13. All the troops were wearing helmets and new flak jackets that General Thuan had gotten for them. (We advisors refused the flak jackets and continued to wear our berets.)

Shortly after 1300 a couple of AK rounds ricocheted off the lead tank and we jumped down from the vehicles. The firing had come from a tree line about 300 meters across an open field. The tanks and tracks came on line and began pouring machine gun rounds into the trees. It might only be one VC sniper, but it also could be a whole damn battalion, so we weren't taking any chances. As the tanks moved forward, groups of Recon soldiers followed close behind, using them as protection. When we got within effective RPG range, the troops deployed and charged the tree line.

There was a deep ditch just inside the trees, and the vehicles had to halt while the Recon soldiers worked their way down and across it. Hien and the squadron commander coordinated on the radio, and some of the armor turned to find a way around the ditch. The rest of the vehicles stayed where they were, in case the Recon needed supporting fire.

We charged up a tree-covered hill, trying to get to the far edge of the woods before Charlie got away. The flak jackets were a lot heavier now that the troops weren't riding on the vehicles, so I got to the edge of the next open field well ahead of the others. I saw a man running across the dry rice paddies, so I dropped to one knee and brought my rifle up to fire. I was breathing heavy, but I could see through the scope that the guy was just an unarmed peasant. As the troops got even with me, I yelled, "Thoi Ban!" [Cease fire!]

In all the excitement a couple of soldiers misunderstood my pronunciation and apparently thought that I had said "Toi" (meaning "I"), instead of "Thoi" (which meant "cease"). In an instant they opened fire, and the peasant went down in a heap.

It took a couple of seconds before I got the firing stopped. The man was about 250 meters away, and the rifle he appeared to be carrying was actually a garden hoe. The soldiers couldn't see that, and if it hadn't been for my rifle scope, I would have shot the guy too.

We ran to where he lay and found that he was an old man, with a bullet through his arm and another through his leg. When the soldiers saw that they had shot an innocent civilian they were upset, but I reminded them that we had just been fired at by a VC sniper. We thought we were chasing him when we spotted the peasant.

Fortunately, although the old guy's wounds were serious, they were not fatal. While a medic bandaged him, I radioed Division and requested a chopper to evacuate him to the hospital. When he found that we weren't going to kill him, he was very grateful for the medical attention—despite the fact that we'd just put two holes in him. Hien apologized, and to make him feel a little better, he stuffed several 100-piaster notes in his hand. The guy was so happy with the money that I thought he might invite us to shoot him again—for a price. The armored vehicles showed up just about the time the chopper was lifting off. Although the old man was in a lot of pain, he waved to us as the helicopter gained altitude.

The sniper was long gone, so we climbed back on board the vehicles and drove on toward Highway 13. Captain Egan and I compared notes and were both pleased to see how well his armor and my infantry were working together.

The trip up to Phu Cuong was uneventful, and we began to debate whether we should roar through the center of town or bypass it to the east. We were about to vote, when suddenly we started taking machine gun fire from the edge of the city, not far from where we had killed the VC with the 250-pound bomb.

This time the VC weren't playing games. They were putting out a lot of fire from a pottery factory to our left-front, a group of concrete buildings straight ahead, and from some hootches to our right. Different tanks took different targets, and I watched as they returned fire with their 76mm main guns.

The first group of VC to be silenced was in the pottery factory. After a couple of rounds, they were either blown away or took off to their rear. I spotted what I thought was movement in a building directly ahead. I yelled to Egan and told him to have his tanks put a couple of rounds into the yellow building just to the left of a large billboard. Egan passed the word to the lead tank and it fired an HE round into the building, right above the front door. After the smoke cleared, we didn't get any more VC fire from that direction.

The main VC position was about 200 meters to our right. The VC appeared to have a couple of platoons dug in around the hootches and trees at the edge of the town. Charlie had picked a good ambush site. The road passed through a large cemetery, and the graves prevented the vehicles from moving off the blacktop or easily turning around.

The more modest graves were high mounds of dirt, surrounded by concrete or laterite rock retaining walls. The richer Vietnamese had three- or four-foot-high concrete tombs with decorative Chinese characters. Because they were so crowded together and solidly built, the graves formed an antitank obstacle as effective as any specially designed for that purpose.

Because of the way the VC were set up, it was apparent they had expected to wipe out a vehicle convoy. Their positions in the pottery factory and building would halt the convoy next to the cemetery, and the graves would prevent any armored vehicles from maneuvering against their main position. While the convoy commander would be arguing for permission to fire artillery into the edge of the town, the VC could destroy the column at their leisure.

Charlie's plan was good, but it didn't consider the possibility that the entire column would be tanks and tracks—and that it would be carrying over 100 aggressive combat soldiers. While the cemetery prevented vehicular movement, it did provide great cover for the Recon soldiers to attack the VC's ambush. With the tanks firing over our heads, we moved into the graveyard and started toward the VC. I was firing around the edge of a tomb when my rifle bolt locked open, forcing me to change magazines. As I reached for a full one, I realized that nobody else was firing—not the tanks, not the Recon, not even the VC—it was completely silent.

I looked to my left and saw Hien and his radio operator standing in full view of the VC. Beyond him other soldiers were beginning to stand up. I had absolutely no idea what the hell was going on. I moved over next to Hien and asked him why everyone had stopped firing. He used a Vietnamese word that I didn't know, and then he pointed toward the VC's positions. Some weird-looking guy was dancing along the edge of the graveyard between us and the Viet Cong. He had two long bamboo poles tied upright on his back. Attached to the poles and his waist were two large red-cloth banners covered with Chinese characters, and they gave him the appearance of a man wearing angel wings. In one hand he held what looked like a carved dragon, and in the other, a bundle of smoking josh sticks. He had a shrill whistle in his mouth, which he blew repeatedly as he danced around, first toward us and then back in the direction of the VC.

I looked at Hien and the others, expecting to see some sort of emotion, but they just stood there with bland looks on their faces. I tried several Vietnamese words, but my vocabulary didn't include anything for this apparition. Hien kept looking at the dancing man and repeated the same word that he had used before. Still confused, I looked around and saw that no one was taking their eyes off the guy, but nobody seemed to be frightened or upset. I looked over to the VC's position and could make out several Viet Cong doing the same thing—they were just standing there, watching this man leaping in the air, waving his dragon and joss sticks, and blowing his whistle.

It took the guy five or six minutes to dance his way across our front, and then he disappeared into the trees on our left. As soon as he was out of sight, everyone got back down, and as if on command, they opened fire again. It was the damnedest thing I'd ever seen.

With the armor's supporting firepower and our assault through the graveyard, we overran the VC after only a few minutes. We killed and captured some, but most escaped into the town and disappeared. Our own casualties were comparatively light, and I had to admit that the flak jackets had come in handy.

When we got back to Lam Son, I spent an hour with Hien and several other Vietnamese trying to find out what the dancing man was all about. The only thing they could tell me was that he was some sort of guy who could call spirits—not a priest or a geomancer, but something like a shaman or witch doctor. Hien said that everyone had stopped shooting because it didn't matter whether you were Buddhist or Catholic; you didn't mess around with spirits or witch doctors. That made sense.

18 ONCE AGAIN

Although still confused about the witch doctor or whatever he was, I headed back to Gosney Compound to clean my rifle and myself. Armor was great for its firepower, and riding beat humping a rucksack, but traveling this way sure covered everything with dust and dirt.

First things first. I started cleaning my rifle before I took a shower. I had the rifle broken down and spread all over the bunk when a strange man walked into the hootch. Since I was the only person there, he came over, flashed an identification card in my face, and introduced himself as a military police Criminal Investigation Division (CID) investigator. "I'm with the 1st Infantry Division, and I'm investigating how some of the 5th ARVN Division's soldiers were able to get M-16s, M-14s, and M-60 machine guns."

If he thought that I was going to get nervous or uptight, or that I would shit in my pants, he was wrong. I kept on cleaning my rifle and asked him why he was talking to me. Why wasn't he talking to the division senior advisor or somebody important?

"I'm just smelling around, hoping that I can come up with some answers without battling the chain of command."

"Shit, man, I'm only a lowly, goddamn lieutenant who bunks in this compound whenever I'm not in the bush—and that ain't often."

He laughed, and I knew that I had scored some points, because despite the fact that he wasn't wearing any rank on his uniform, I knew he was either a sergeant or a warrant officer. Those kinds of guys liked it if you admitted that you were just a dumb lieutenant. It was perverse logic, but saying that you didn't know anything proved that you really did.

"How do you figure these ARVN are getting the new weapons?"

I stood up and went through the motions of moving the rifle parts to the end of the bunk so I could get comfortable. In the process I covertly pulled the camouflaged poncho liner so that it touched the floor on the side of the bunk where he was standing. That concealed the half-dozen M-16s and one M-60 I'd put under the bunk to get them out of my way. Glen and Tony had just picked them up from the very people this investigator was trying to find.

Sitting back down, I told him that I didn't know about most ARVNs, but my people had been capturing quite a few from the VC. "Charlie must have done a job on those suckers in Saigon, because even the VC local-force units have M-16s now. It's a son of a bitch going up against AKs and M-16s with only M-1s, carbines, and BARs."

"I see that *you* have an M-16. Aren't advisors issued carbines and .45s?"

"Yup, I got this baby off a dead VC—who probably got it off a dead GI from your division."

The guy got up, put his notebook away, and shook my hand. "If you get wind of any black market dealing in weapons, we'd sure appreciate it if you'd let us know."

"Sure thing; but as I said, I'm just a grunt, and nobody tells me anything."

I had a hard time holding in my laughter until the guy was out of earshot. Even if he had caught me, I didn't figure anyone would do anything. I wasn't personally profiting from the illicit weapons traffic—we were giving them to the Recon so it could kill more VC. Besides, what could they do to me? As the old saying went: "They couldn't send me to Vietnam."

After I washed up, I went to Major Clark and Colonel Sonstelie and told them about my visitor. Both of them said for me not to worry. Laughing, I said that I wasn't, but that they ought to get on somebody's case about American investigators sneaking around a Vietnamese headquarters.

The next day was supposed to be another day off, but I hadn't even gotten my heart started beating before the phone rang. A Recondo patrol was in trouble, and I was wanted in the TOC immediately. Just after dark on the preceding night, a chopper had inserted one of our eight-man, long-range reconnaissance patrols. The Recondos were supposed to move out of the LZ and be at least a couple of kilometers away by dawn. Unfortunately, the chopper had inserted them near a pro-VC village that just happened to be holding a whole company of visiting Viet Cong. The villagers had begun beating pots and pans, and the VC had surrounded the big open field the patrol had landed in. Our people had hidden in a big bomb crater, and the VC hadn't been able to find them in the dark. Now it was light and Charlie was coming to get them.

The area was well beyond the range of a PRC-25, so the patrol couldn't talk directly to Division. At prearranged times one of Hooker's Bird Dogs was supposed to fly overhead with a bilingual ARVN soldier in the back seat. The soldier would contact the patrol and relay messages between it and Division. It wasn't until the first flight this morning that Division found out that the patrol was in deep trouble.

It's pretty tricky trying to extract a patrol that has been compromised—but even more so if the patrol is sitting out in the middle of an open field, surrounded by VC. I asked Division what it could give me, but the only thing they had was a single slick. A pair of gunships were on their way to Lam Son, but no airstrikes were available until later in the morning.

I grabbed Hooker and asked him to go with me to the chopper pad. When we got there, I gathered the slick crew together and told them that I had a patrol in trouble. I explained the situation, including the fact that the field was surrounded by a VC company. I concluded by telling them that if I didn't get the patrol out right away, I'd lose eight good men. Since this was going to be a hairy operation, I asked them if they would be willing to fly the mission for me. It only took them about five seconds to volunteer.

"Oh, there's just one more thing—the patrol is dressed in black pajamas and is carrying AK-47s."

The pilot threw up his hands, but before he could say anything, I told him that I could recognize the patrol, because its men were wearing red and black scarves. One of the door gunners would have to remain behind so we wouldn't overload the chopper, and I would take the man's place. That way, if we started to land on a group of real VC, I could blast them with the door gun. They were less enthusiastic than before, but they still agreed to go. (Thank God for teenage warrant officer pilots.)

I asked Hooker to handle the artillery and gunships for me, and I showed him where I wanted him to put them. The artillery would fire along one edge of the field, and the guns would put rockets along the other edge. Our slick would come in between and parallel to the impact areas. I hoped we could get in and out fast enough to avoid getting shot down.

Everybody agreed to the plan, so I put on the door-gunner's chest armor and flight helmet and climbed into the right door. Hooker and the gunships took off first, and we trailed them a few seconds later. We only had to go about fifteen kilometers, so the slick had to circle well south of the field until Hooker got the artillery where we wanted it. Then when the gunships were rocketing the far side of the field, we headed in fast and low.

There were several large bomb craters in the field, but only one had eight black figures in it. The pilot spotted them first, and I leaned out the door to verify that they were my people. It wasn't hard to tell, because they were frantically waving their arms.

The chopper hit hard and bounced a couple of times before coming to rest at the edge of the crater. In a flash, eight Recondos were crawling in the door. With my left hand I was grabbing whatever I could to pull them in, while firing the machine gun with my right hand. To hell with aiming—I just wanted to keep the VC's heads down until we could get everybody on board.

We had six all the way in the chopper and two hanging out the door, when the pilot pulled pitch and we were airborne. It was a great feeling when I saw the edge of the field disappear beneath the helicopter. The six Recondos inside the aircraft pulled their two buddies into the cargo compartment, and eight very happy Vietnamese soldiers began cheering.

There was a lot of hand-shaking and back-slapping when we landed back at Lam Son. The Recondos offered the crew their AK-47s and anything else the Americans wanted. After the pilots and door gunner had had a chance to admire their war souvenirs, we checked the helicopter. It had about a dozen holes in it, and hydraulic fluid was leaking on the ground. The pilots started the engines, and after checking the instruments, they decided that they could get the thing next door to Phu Loi before it fell apart on them. We waved as they took off, and the Recondos and I walked back into the Lam Son compound.

Tony and I debriefed the patrol and got the story on what had happened. Their chopper had made several false insertions, both before and after inserting the patrol, to confuse the VC on where they were actually going in. Immediately after the helicopter departed the real LZ, the villagers began banging pots and pans to alert the VC that somebody had been dropped in the field. The Recondos said that the peasants had helped the Viet Cong search for the patrol by shining flashlights and kerosene lanterns. As the Recondos tried to get away, more lights appeared on the far side of the field, and they knew their escape route had been closed.

Since only one chopper had landed, the VC knew that there couldn't be more than ten men in the patrol. That wasn't enough to worry them, so small groups of VC and villagers kept crisscrossing the field, searching for the Recondos. Although they were nearly discovered several times, the high grass and darkness saved their asses.

After several hours of searching, the VC pulled back to the edge of the field and began firing M-79s and dropping mortar rounds into it. The Recondos probably would have been killed or captured had it not been for the bomb crater they'd stumbled into. They kept trying to contact somebody on their radio but couldn't get anyone who spoke Vietnamese. For some reason the VC didn't come after them at first-light, and were only just beginning to sweep the field when the artillery started falling. The artillery could not have fired indefinitely, so if we hadn't come in with the rescue chopper, the Recondos would have been dead meat.

We'd had a lot of troubles with this particular village before. Each time we or the Americans went near it, we lost people from ambushes, snipers, and booby traps. Province headquarters classified this village as Viet Cong—not just pro-VC—and always approved artillery and air whenever it was needed. It wasn't as big as Phu Hoa Dong, but it certainly was as bad.

By the time the Recondos were done telling their story, I was angry. The Vietnamese government had relocated other VC villages, and I wondered how long we'd have to put up with this one before it finally went in and cleaned it out. Relocation was certainly no solution to winning the war, but since the peasants in this village were never going to come around to the "hearts and minds" shit, we just ought to move all the villagers to some place where we could watch them.

Hooker had been listening to the debriefing, and I nodded for him to join me outside. "I've got an idea. Do you want to help me harass those fuckers a little?"

After he agreed, I told him to get his airplane ready—I'd meet him in the tie-down area in fifteen minutes. I grabbed a jeep and drove over to the Recondo compound. The Recondos lived in one of the old POW barracks, and Division stored captured VC weapons and ammunition in another. I had never been very keen on storing large amounts of ammunition and explosives in a building, but nobody seemed to be concerned about it. I did have to admit that it was a handy arrangement when we needed ammunition for the Recondos' AK-47s.

I loaded up a box of stuff and drove back to the airstrip. Hooker was standing by the airplane and did a double take when he saw that I had eight 82mm Chicom mortar rounds. "What're you planning to do—bomb the village?"

"You got it. That's exactly what we're going to do."

"Don't mortar rounds have to be fired before their fuses are armed?"

"Ours have set-back arming, but Chicom rounds don't."

Before he'd climb in the airplane, Hooker made me go through the entire procedure with him. I'd have to hold the mortar round out the window with one hand and screw the fuze in with the other. After the fuze was seated, I'd have to unscrew the small metal cap protecting the detonator. There was no way Hooker was going to let me put a round together in the back seat of the Bird Dog, but if I held it out the window, it would drop free if I fumbled it, and it wouldn't blow us out of the sky.

When we got over the village, I stuck a round out the window and screwed in the fuze. When I had removed the cap, Hooker lined up and I released the thing. As soon as I yelled, "Dropped!" he banked around, and we watched the thing explode near the edge of the village. I wasn't going to risk being put in jail for killing "innocent civilians." I knew that when the mortar round exploded outside the village, the people would quickly haul ass into their bunkers. I only wanted to harass them; I didn't want to kill any little kids. During our next seven passes, we didn't see anybody, but we did blast a few water buffalo and a bunch of pigs. After I had dropped the last one, we flew low over the fields and I zapped some more water buffalo with my M-16. Then we turned and headed for home.

It wasn't until a few days later that I found out what had happened after we'd left. Just as I knew they would, the villagers had complained to the South Vietnamese government district chief, saying that an American airplane had bombed their village. A very unsympathetic group of ARVN and U.S. officers had investigated and found Chicom 82mm mortar fin assemblies in the craters. I understood that the village chief had sputtered, but he couldn't come up with a response when they told him that the village must have been mortared by the Viet Cong.

The next day we were on our way back into the Lam Son Secret Zone again. We were all getting thoroughly fed up with this place, and I even considered not going with the company. Of course, it was my unit, so I couldn't very well let it go without me. Since Hien didn't have any choice, I supposed that I didn't have one either.

There was no use wasting the others' time, so I told Chuck and Glen that I'd only take Collins. They could go and scrounge some more weapons and pick up other supplies that the two companies could use. Sergeant Collins wasn't upset about going; he was really into this combat stuff. He had changed; he was no longer the young inexperienced perimeter guard he'd once been. We still had to keep an eye on him to make certain that he didn't do something to get blown away, but he was definitely turning out to be a good addition to the team.

On our way into the area, we had to duck some ricochet rounds from the division rifle range, but that was the only excitement we had until shortly before noon. We had just eaten and were moving again when the point team opened up on something. Everyone took cover behind the trees, and Hien radioed the lead platoon to find out what was happening. Before he could get an answer, there was a big explosion, and a nearby hootch blew to pieces. Whatever it was, it wasn't an RPG; the blast was too powerful. I hunkered down as low as I could as dirt and pieces of wood rained down around me. I was scared; I'd learned a valuable lesson at Phu Hoa Dong—*the VC really could kill me.*

Everything was thoroughly confused. There were bullets cracking by from all directions, and the Recon was firing back at the unseen enemy. The first few minutes of a firefight are always that way, but once the surprise wears off, you can usually begin to piece together what's happening. Today I had no idea what we had walked into. There was so much firing that I couldn't figure out where the VC were—or even make a rough estimate of how many people they had. I did know that I was exposed and had to get to a better position.

Taking a big breath, I jumped to my feet and ran toward a thatched hootch about fifteen meters away. Despite the noise and my heavy breathing, I could hear bullets snapping by me as I ran. At least one of the bastards had spotted me and was trying his damndest to shoot me.

I threw myself behind the hootch, but before I could do anything, it blew up! Dazed, I jumped up again and raced to the next one. *Damn!* Although I couldn't see them, the VC seemed to be everywhere—and they all seemed to be trying to kill just me.

My heart was pounding, I was gasping for breath, and I was scared shitless. I frantically looked around, but I couldn't see anyone—not even my own troops. I felt the terror that most soldiers feel at some time, when they think that they are all alone and surrounded by the enemy. I moved to the side of the hootch and cautiously looked around, but I still couldn't tell where my people were. Pressing myself against the mud and straw wall, I eased my way to the other side and peered around the corner. I was on the verge of panic when I saw a VC grenade sailing toward me. Everything was in slow motion—the grenade slowly got bigger and bigger as I tried to make my legs move me to the other side of the hootch.

Boom! The grenade exploded at the corner of the house, and I could feel a blast of dirt and small rockets hit me. The force of the explosion threw me on my ass, and I hit the ground hard. Stunned, I lay there for several seconds, wondering how badly I had been hit.

Slowly the fog disappeared and I began to feel my body, searching for wounds or blood. Except for the pain in my shoulder from the fall, I didn't feel injured. My ears were ringing and I had dirt in my face and eyes, but I seemed to be okay.

Slowly I got to my feet and brushed the dirt out of my eyes. Surprisingly, I wasn't scared any more—but I was mad. No low-life, VC son of a bitch was going to blow me up and get away with it! I got my own grenade from my shoulder harness and pulled the pin. I sneaked to the corner of the hootch and heaved it in the direction the VC grenade had come from. Before it exploded, I was reaching for another one. I threw the second grenade in the same area. When it exploded, I stuck my rifle around the corner and sprayed a full magazine in the bushes.

Figuring that I had gotten the bastard, I moved back to the other corner to see if I could see any of my troops. This time I saw a line of Recon soldiers moving through the bushes, firing from the hip and shoulder. The nearest one was only about ten or fifteen meters away, and apparently he had been there all the time I had thought I'd been alone.

A wave of relief swept over me as I joined the assault. I never saw anything the whole time we were moving forward, but I fired almost a full bandoleer of magazines before we halted. The signal went out to cease fire, and a silence fell over the area. There was sporadic shooting and bursts of fire, but we had evidently killed all the VC who had stayed to fight.

I collapsed against a tree and rested, while the troops searched the bushes. Every time they found a dead Charlie, they stripped the body of its weapon and equipment and brought them back to the command group. I only saw four or five bodies, but Hien said that we had killed eleven. We had seven AKs and three SKS carbines, so if anything, he was probably understating how many we had gotten.

As I stood up, a soldier looked at me and pointed to my legs. I looked down and saw that my fatigue pants were covered with blood. I took off my web gear and pulled down my trousers. I had a half-dozen small holes oozing blood—the VC grenade had gotten me and I hadn't realized it. One of the medics cleaned the wounds, and after pouring iodine on them, he bandaged my legs.

Only three of us were wounded, and we were all able to walk, so I didn't call in a Dust Off. On the walk back to Lam Son, I did some serious thinking. I had been in the bush for almost eleven months and wondered if it was time to come in. I no longer had an illusion of invincibility, but at last I was being tested. Until Phu Hoa Dong, I had regarded combat almost as a lark. Certainly I mourned my dead and wounded friends, but as far as my own safety was concerned, combat had just been exciting. Now I knew that I could fit in a body bag just like everyone else.

It took almost two hours to reach the compound, but in that time I came to a decision. I'd stay to the very end of my tour, provided the wounds didn't become infected or the shrapnel didn't have to be removed by surgery. Actually, the decision was the easy part. I'd have to conceal that I'd been wounded a second time—that would be the hard part—or Sonstelie and Clark would yank me out of the field and stick me in a desk job. I told Hien what would happen if they found out, and I asked him to pass it out to the troops to keep quiet. I didn't even want them to tell the other guys on the team. I was afraid that they might let it slip, and then I'd be cooling my heels while they were out killing VC.

Before I left the Recon compound, I had a medic give me a bottle of iodine, a handful of gauze bandages, and some tape. Whatever medical attention I might need, I'd either take care of it myself or have a Recon medic do it.

257

19 FUN AND GAMES

I was sitting in the G-2 office planning an Eagle Flight, when Major Clark said that he was going to make me famous. "There's a CBS television reporter coming up from Saigon this afternoon, and he wants to do a story on an ARVN combat unit. Colonel Sonstelie wants you to take the guy and his cameraman on your operation tomorrow."

Many of us in the military hated reporters, because they always seemed to be stabbing us in the back and looking for something negative to write or televise. I wanted nothing to do with the press, and I argued that Division should make some other unit take the guy. Despite my protests Major Clark stood firm and said the decision had already been made.

I had been allocated two light-fire teams and a full lift helicopter company. With that many slicks I could take most of the Recon and bring in both of the standby Recondo platoons if we needed them. I had been planning to work the area on the east side of the Thi Tinh River, but I decided to change the AO. With a reporter along, I wanted the best chance of VC contact, and that was in the Iron

258

Triangle. If he wanted to see how well an ARVN unit could fight—we'd show him.

That evening Colonel Sonstelie introduced me to the reporter and his cameraman. I had hoped for some well-known sucker like Walter Cronkite, but I didn't recognize this guy. He looked like a jerk with his jungle safari jacket and khaki pants, and when he shook my hand, he acted as though he expected me to ask for his autograph. He did offer to buy me a drink, and then he asked where we were going the next day. I swallowed a mouthful of beer and said, "The Iron Triangle."

The way he reacted, you'd think that I had kicked him in the balls. "We're going into the Triangle *with only a hundred men*? The Americans don't usually go in there with less than a battalion."

"Yeah, but you do want some combat footage, don't you?"

The next morning I picked him up at the mess hall and drove to the airstrip. I was a little surprised he was there. After his reaction the night before, I'd thought that he probably would have been called back to Saigon for some important story. I introduced him to Hien and Lieutenant Dac, Hien's executive officer, and I could see that he was concerned. Between Hien's schoolteacher image and Dac looking like he was fourteen or fifteen years old, the reporter was worried that he was going on some sort of Children's Crusade. His attitude did change when I walked him around the troops. As usual, the Recon soldiers looked tough and were armed to the teeth.

It was a little crowded in the command and control bird, but the reporter, his cameraman, Hien, Collins, and I all jammed in the back. When we were buckled in, the chopper and two of the gunships took off. We flew northwest and in a few minutes crossed the Thi Tinh River into the Triangle. The bird was holding level at about two thousand feet of altitude, and that was way too high to see anything on the ground. I pushed the intercom button and told the pilot to take us down so I could find someone or something to put the Recon onto. As had sometimes happened before with other aviation companies, the commander gave me a ration of shit about risking a million dollar helicopter. He suggested that we have a gunship get on the deck to see what it could find. (Young warrant officers would usually go wherever you wanted, but captains and majors sometimes had to be hit over the head. It seemed like the higher the rank, the higher the pilot wanted to fly.)

"Listen up, Major. No Victor Charlie in his right mind is going to shoot at a gunship, and your million dollar airplane doesn't mean shit to me. Get this thing down where I can see the ground!"

He started to say something but apparently changed his mind. He stuck his hand out and jerked his thumb downward, indicating to the copilot to take us down. We dropped about a thousand feet and leveled off. We were still too high for me to see anything, so I got up, leaned over the major's shoulder, and pushed the intercom button. "Do you see the number five at the bottom of your altimeter dial? I want you to peg the altimeter's big hand on that five and the little hand on the zero [500 feet]. That's the *highest* I want this damn aircraft!"

I thought that he might turn around and punch me in the nose, but after a couple of seconds, the bottom dropped out and we dove toward the ground. (*I think I pissed him off.*) He jerked the bird level and the cameraman almost lost his camera out the open door.

We flew around for about twenty minutes, but we couldn't get anyone to shoot at us. I was getting pretty frustrated, when I spotted a couple of old bunkers under some high bushes near a large open area. I had one of the gunships put a couple of rockets on them but couldn't tell if there was anybody in them.

We went on searching, but we came up dry. The VC seemed to be taking the day off. Since we had to do something for the reporter, I decided to go ahead and put the Recon in near the bunkers we'd seen. In a few minutes the flight leader called and said that the slicks were inbound and would be ready for insertion in five minutes. I told the commander to have two of the gunships prep the LZ while we gave the slicks final instructions.

The reporter told the cameraman to film the air assault, so we had to wait until the Recon was on the ground before we landed with the C&C bird. On the final approach I told the aviation major that after he dropped us off, he was to take the birds back to Lam Son and wait for further orders. Out of the corner of my eye, I saw him give me the finger as I ran out of the rotor wash.

The slicks and two of the gunships took off, leaving the Recon in the field while the remaining light-fire team orbited overhead. Hien gave the command and the troops pushed into the tree line toward the bunkers. The tide had just gone out, so the field was muddy, with pools of standing water. As I had hoped, the reporter got his nice safari suit dirty.

The bunkers were empty, so we moved farther west to see what else we could find. We'd been moving for about twenty minutes without finding anything, and I was beginning to think that we were going to come up empty-handed. Just then, a couple of AK rounds cracked overhead. It was probably only a sniper, but anything was better than nothing. Hien signaled, and the troops began firing and moving toward where the rounds had come from.

We started taking more VC fire, but it still seemed that we probably only had a Viet Cong squad. It wasn't bothering the troops that much, but the reporter was scared shitless. He kept going to ground, but when he realized that we were leaving him behind, he'd run like hell to catch up. His cameraman seemed to be taking it all in stride and kept filming as we moved forward.

I had hoped to find something bigger, but I only had a two-hour window for using the airstrike we'd been given. I called for the FAC and used the gunships to keep the VC pinned down until he arrived.

I wished that the troops would have acted more convincingly, but they just leaned against their rucksacks and smoked cigarettes, while the F-4s blasted the VC. They knew that this was a big show for the American television reporter, but they couldn't have cared less. Even Hien looked bored as the bombs exploded and the napalm flamed.

The reporter didn't seem to notice, because he crawled up to me and stuck a microphone in my face. While the cameraman stood filming, he asked me, "How many Viet Cong are we up against?"

"Oh, only about five or six men."

"Cut! You're going to have to say there are more than that or they'll never show this on the nightly news. Let's try again."

"Okay, Lieutenant Parrish, how many VC are in that tree line?"

"A five- or six-man squad."

After telling his cameraman to cut again, he spent five minutes trying to convince me that I needed to say that we were up against at least a company of NVA.

"Let's try again. How many NVA do you think are in that bunker complex?"

"What bunker complex?! We've only got about a squad of VC, and they're hiding in some bushes. The damn airstrike is for your benefit. We sure as hell don't need it to take care of a lousy squad."

"*Cut!*"

A very pissed off reporter crawled a few meters away and lit a cigarette.

When the airstrike was finished, we moved forward and found two VC, who had been burned to a crisp. That was all we found, but the reporter had his assistant film the smoking bodies from every angle. I was certain that he'd splice the airstrike film with the fire-and-movement segment. Then he'd add the burned bodies and manufacture a fantastic battle scene for the folks back home.

Since we had exhausted this area, I called for the choppers to come in and pick us up. Every time I held the radio handset to my mouth, the reporter stuck his microphone in my face, but each time I spoke I made certain to include at least one word forbidden on television. The reporter kept asking me to use cleaner language, but I always dropped a "fuck" or a "motherfucker" in each transmission. (I wondered what he could do with *that* tape.)

When we pulled out of the LZ, the reporter leaned back in his seat, checked his watch, and asked if we were headed back to Lam Son.

"Nope. We've got the choppers all day so we're going to use them."

I inserted us in another area, but this time we didn't even find a sniper. After two hours of wading around in the mud, we pulled out, and I told the pilot to take us about eight kilometers north of Lam Son. There probably wouldn't be anything in that area, but I had one more dirty trick up my sleeve. After a very tired reporter stumbled off the chopper and the birds flew away, he asked me what I planned to do now.

"I've released the choppers for the day, and we're going to walk home, sweeping the area between here and Lam Son."

"Shit! How far is it?"

"Oh, only about eight or nine kilometers. If we don't run into anything, we should be back at division headquarters just before dark."

"Christ! What happens if we find something?"

"Then we stay out all night."

The cameraman was carrying a big heavy camera, while the reporter had a small tape recorder in a shoulder bag. Despite the equipment, the cameraman didn't seem to be having any trouble, but his reporter buddy was completely exhausted by the time we reached Lam Son. When I left them at the gate at Gosney Compound, the reporter didn't even bother to shake my hand. The cameraman did and gave me a big wink.

The next day Colonel Sonstelie got hold of me and said that it looked like something was going to happen. MACV and our own intelligence indicated that the VC might be getting ready to repeat the Tet business. General Thuan was convinced that this time Charlie would hit Lam Son with a big attack. "The general and I would like you to take a look at our defenses and come up with some recommendations on how we can strengthen them."

There wasn't much to look at—some barbed wire, some claymores, and an old World War II Japanese minefield. I walked the berm, made a few notes, and went back to Sonstelie. I suggested that they reinforce the bunkers, put out more claymores, and relocate the automatic weapons every night, but my main recommendation involved the minefield. The Japanese mines had been in the ground for more than twenty years. Obviously they weren't worth a damn any more, but they could make putting in new mines very dangerous.

I drew a couple of sketches and showed Sonstelie that his men would have to penetrate the minefield with bangalore torpedoes and bury some new mines. They couldn't use mine detecting equipment because no one knew what kind of mines the Japanese had used, and the barbed wire made metal sensing devices virtually useless. I showed him that by starting at the road that ran through the field and out to the airstrip, they could blast several paths parallel to the berm. Then they could put in new mines.

He liked the plan and said that he wanted me to start on the Gosney Compound part of the minefield. I protested that I wasn't an engineer officer and that there were other better-qualified people to do the job. He nodded but said the ARVN engineers and their advisors were going to be busy with the division's compound. They'd get me whatever I needed, but I would have to do the job with only my team.

The next day a truck drove into the compound with a load of bangalore torpedoes and the stuff that I needed to make *foo-gas*— homemade napalm bombs. The bangalore is a light metal pipe which is a couple of inches in diameter and about six feet long. It's loaded with about five pounds of explosive. Each end of the pipe has a metal clip so the tubes can be connected. The engineers begin by pushing a section through the barbed wire. When they get to the end, they clip on a second section and continue pushing. They can use as many six-feet sections as they need, or as many as they can push. When the thing is in place, they roll out electrical wire, attach it to a detonator in the end, and fire the sucker off. The explosion detonates mines and blows a gap through the wire.

I took the other guys and guided the truck to the road through the barbed wire. Theoretically the procedure is simple, but it takes big balls to push a pipe filled with explosive through a minefield. If I set off a mine while sliding the bangalore along the ground, there wouldn't be enough left of me to fill a small plastic bag. The mine wouldn't just blow, it would also detonate the bangalore in my hands.

The ARVN soldiers broke open the wooden boxes and handed me the first section. I screwed the rounded nose cap on the end so the tube wouldn't hang up, and then I carefully slid the section under the wire. It went smoothly and I asked the soldiers for another section. At first I was scared, but then I decided that if the Japanese mines had detonating prongs, they must have rusted away years before. The bangalore was probably not heavy enough to set off a pressure mine, so if I were careful, I'd be okay.

I kept adding sections until I was having a hard time pushing the thing. I had the soldiers unload six or eight more sections and then ordered them down from the truck to help me push. I estimated that it was close to 100 feet long and had about sixteen sections. I decided that we'd better stop and set this one off before going on.

A large audience was standing on the berm and watching, so I yelled that I was going to blow it. Starting from the gap in the berm, I reeled out the electrical wire and attached it to the end of the bangalore. I went back to take cover and had Tony make certain that everyone was behind the berm. When he gave me the high sign, I yelled "Fire in the hole" three times and turned the handle on the electrical generator.

The blast surprised the hell out of me—it was absolutely awesome! I'd never seen a bangalore fired before and didn't know what to expect, but I certainly didn't expect that big a blast. The shock wave reminded me of the films I had seen of the atomic tests in Nevada. First a wall of dust and dirt blew into the compound, and then a second later it blew back toward the minefield as the air rushed to fill the vacuum pocket created by the explosion.

I had just started walking back to the blast area when I heard the alert siren start up in the compound. Somebody apparently thought that Gosney had taken a VC rocket and was cranking the alarm. The people climbed back on the berm and a loud cheer went up. Bowing to my audience, I motioned for the ARVN soldiers to drive the truck back to the work area.

We were carrying bangalore sections down the blown path, when somebody on the berm began yelling and waving his arms. I walked back to the road and saw Lieutenant Colonel Sakas marching toward me. Colonel Sakas was Sonstelie's deputy and not one of my favorite persons. As far as I was concerned, he was just a rear-echelon type, and I had as little to do with him as I could.

Before I could open my mouth, Sakas angrily began raising hell. He didn't know what I thought I was doing, but I had scared the shit out of everyone. Not only that, the blast had broken every lightbulb and piece of glass in the compound. It had even blown down a plywood wall in the NCO club. Ordering me to keep my mouth shut and follow him, he took me on a tour of the devastation my bangalore had caused—it was impressive. I had no idea that it had that much power. The *pièce de résistance* was the fact that the blast had broken General Thuan's flush toilet.

Sakas marched me to Colonel Sonstelie's office, still refusing to allow me to speak. When we got there, Sonstelie was replacing some things that had been blown off his desk. When Sakas finished detailing all the damage I had done, Sonstelie asked me what had happened. I told him that I had been doing what he had ordered me to do—I was clearing a path through the minefield.

"How many bangalore sections did you use?"

"I don't know, sir, probably fifteen or sixteen."

"Jesus! Didn't you know you can't fire that many at once?"

"No sir. Remember, I told you I was an infantryman, not a goddamn engineer officer!"

Sakas didn't see anything funny about any of it, but Sonstelie started laughing. I had warned him, but it had never occurred to him that I might blow *seventy-five or eighty pounds of explosive* all at once. "How much more do you have to do?"

"Probably another 100 meters, sir."

"Okay, but from now on, don't use more that three or four sections—and warn the TOC before you blow them."

I went back to the minefield and continued with the project. There was some booing when the next blast went off, because it wasn't nearly as impressive as the first one. It took much longer to clear the rest of the path because of the limit on the number of bangalore sections I could use, but by late afternoon we had finished.

The next morning I went to see Colonel Sonstelie, but he had flown to a meeting at Bien Hoa and wasn't scheduled to return until late in the day. That meant that I had to deal with Colonel Sakas. I told him that I wanted to burn the grass and weeds that had grown up in the minefield. It was so overgrown that it would be almost impossible to detect a VC sapper working his way through the wire.

Sakas was still pissed about the day before and asked me if I had already cleared the burning with Sonstelie. I told him that I hadn't and that's why I was in his office now. Like bureaucrats everywhere, he told me to do *only* what Colonel Sonstelie had approved, nothing more. There was no use arguing with him, so I just shrugged and said, "It's your compound, sir."

Our last big job was to put out barrels of foo-gas. Making foo-gas is a lot of work for a one-shot weapon, but it will put a hurt on a human-wave attack. First you dig a shallow hole about a foot deep. Then you wrap a block of TNT (or C-4) and two white phosphorus hand grenades with detonating cord and stick an electrical blasting cap in the TNT. You put the bundle of explosives in a plastic bag to keep it from getting damp and lay it in the bottom of the hole.

Carefully, you set an open-ended fifty-five-gallon oil drum on top

266

of the charge and sandbag it so it angles about fifteen or twenty degrees toward the enemy. The dirty work starts when you fill the drum with gasoline, add a thickening agent, and stir the mixture until it is the same consistency as gelatin.

When the VC get into your barbed wire and if you haven't been able to stop them with your other weapons, you fire off your foo-gas. The TNT explodes, the white phosphorus ignites the jellied gasoline, and burning napalm is blasted over the attackers. It's a last-ditch thing, but it is really effective, particularly if the VC are bunched up.

We spent a long hot day carrying five-gallon gas cans and stirring the mixture, and we were covered with oil when we finally finished the last foo-gas drum. The engineers or perimeter guards could replace the mines and barbed wire. We had done all that Sonstelie had approved.

The next day it was back to the field. A convoy of trucks picked up the Recon, and we headed down Highway 13 to the southern edge of Binh Duong Province. We'd been ordered to search for any signs that the VC might be moving toward Saigon. We unloaded south of Lai Thieu and began searching west toward the Saigon River. It was only about a kilometer away, so it didn't take long to reach it. We turned south and swept another kilometer or so and then headed back to Highway 13. We'd found nothing, and the most dangerous part of the morning's operation had been our efforts to get 100 men across the highway without being run down by fast-moving cars and trucks.

Despite the fact that it was the dry season, the fields were lush and green from the many small canals that carry water from the river. We had moved about 100 meters down a path from the highway when we came upon a small house with a tile roof and a front porch. Hien halted the company and gave instructions to break for lunch. We were hot, sweaty, and thirsty, and the troops greeted the order with relief.

I began eating a sandwich that I had bought at one of the little pushcart foodstands along the road. (Although the Vietnamese never understood why, American GIs called them "Howard Johnson" stands, after the U.S. restaurant/motel chain.) I was about half done when a couple of soldiers came up to the porch, escorting a young man who was wearing only shorts. They said that they had found him in a nearby canal, acting as though he was repairing a wooden sluice gate. Hien was somewhere checking with his platoon leaders, so I told them to leave the guy with me.

Continuing to eat my sandwich, I asked the man if he lived here. He was surprised that I spoke Vietnamese and nervously answered yes. I took another bite and asked him why he wasn't in the army, since he was the right military age. He explained that he had been in the army but had been discharged after he'd been wounded. He showed me a scar on his arm, but it didn't look serious enough to justify a discharge.

Since I wasn't acting threateningly, he relaxed a little and asked me if I would like some coconut milk to go with my sandwich. I nodded, and after brushing his feet off, he climbed right up the smooth trunk of a nearby palm tree. (I could never figure out how they did that.) After knocking a coconut loose, he shinnied back down. Collins gave him a machete and he expertly whacked off the top of the nut and poked a hole in it. After Collins and I drank some of the milk, the man grinned. He was no longer nervous, and he started acting friendly.

I chatted with the guy and noticed that he spoke with a North Vietnamese accent. For example, he used a sharp "S" when he said "Saigon." In the southern dialect, it would be pronounced with an "Sh," like "Shy-gone." Unfortunately, you couldn't tell the good guys from the bad ones by the dialects they used. In 1954 almost a million North Vietnamese had migrated to the south to get away from the communists. In fact, about half the ARVN officer corps was North Vietnamese.

Apparently the man became concerned that I understood the language, so he told me that his family had come south in 1954, when he was just a small boy.

"How long have you lived in this house?"

"More than ten years, sir."

"Do you ever go to Saigon?"

"No, sir. I have only been to Saigon twice in my whole life."

"Only twice? Saigon is not very far away, is it?"

"No, sir. It's only about fifteen kilometers from here."

Just then Hien and Dac came back, and I offered them some

coconut milk. Seeing that it was almost gone, the man climbed back up the tree and knocked down another one. When he got back on the ground, Hien began to question him. I took another swig of milk and said, "Don't waste your time, Hien. He's a North Vietnamese soldier—probably an officer."

The man's eyes grew large, and he started protesting that he was just a poor peasant. I told him to shut up and then explained it to Hien. "He speaks with a North Vietnamese accent and uses too many big words for a poorly educated peasant."

Hien frowned at little and said, "But there are many northerners in South Vietnam."

"That's true, but this man says he has lived in this house for more than ten years. When I asked him how far it was to Saigon, he said, 'About fifteen kilometers.' "

"What does that prove?"

"A hundred meters down the path is Highway 13. Next to the path there's a concrete kilometer marker that says it's 11 kilometers to Saigon. How could he have lived here ten years and never have noticed what the marker said?"

Hien smiled and gave orders for the man to be tied and blindfolded. When we got back to Lam Son, we turned him over to the G-2. During interrogation he admitted that he was a senior lieutenant in the 9th VC Division.

20 PROMOTION

On 1 April 1968, I was promoted to captain. Before I arrived in Vietnam, it took twenty-four months before you could be promoted from first lieutenant to captain. Now with so many captains being killed and wounded, the army had to speed up promotions to fill all the vacancies.

It cost me a small fortune, but I threw a big party for the Recon and Recondo companies, and the handful of my American friends. To be polite, I invited General Thuan, and to my surprise, he accepted.

I begged and borrowed other people's ration cards until I had enough to buy booze for over 250 people. The Recon put up a couple of banners congratulating me and decorated one of their hootches for the party. After polite speeches by General Thuan, Major Tu, and Lieutenant Hien, we started eating and drinking. The troops were on their best behavior until Thuan and the other "biggies" excused themselves and returned to the division compound—then things got wild.

Most Americans could handle alcohol better than the smaller Vietnamese, but our soldiers were persistent. As at Gene Bolin's party, individual troops would come up and challenge me to chug-a-lug water glasses which were filled with whiskey. But in the past eleven months I had learned more than just how to survive in combat. After every couple of glasses, I would make my way outside, pretending to be going to relieve myself. Then behind the hootch, I'd stick a finger down my throat and puke up the whiskey. After cleaning my face, I'd go back inside and meet the next challenger.

We didn't make it to midnight. By 2300 the compound looked like it had been overrun by the VC, with dead (drunk) bodies lying everywhere. General Thuan knew what would happen, so he had approved rest and recuperation for the following day. Everyone needed it but me. I woke up feeling great and spent some time admiring my new captain's "railroad tracks." Apparently no one ever figured out how I survived the party, and I gained the reputation as a world-class drinker.

That morning I went to the mail room and found a note from an old friend. He had called Washington and found out that I was being assigned to the army's Ranger School as an instructor at the Ranger camp in Florida. Shit! That's all I needed—a year or so wading around the swamps, teaching jungle warfare and Ranger tactics, and then back to Vietnam. I had expected a less demanding job, one where I could have a home life and become civilized again.

I threw the note on the bunk and wandered over to the supply room. Parked outside was a two-and-a-half ton truck; its front end was smashed and its windshield was bullet-riddled. I walked into the building and asked a very shaken young sergeant what had happened. He said that he and another man had been returning from a supply run to Long Binh when they had been ambushed on Highway 13. Without warning, the VC had opened up on them and had just missed hitting the truck with an RPG-7. They ran off the road and crashed into a ditch. The sergeant said that they had crawled out of the cab and into the high grass growing in a large muddy field. While some of the VC ransacked the truck, others began searching for him and his buddy. Scared shitless, they lay in the grass, praying that the VC wouldn't find them.

"Did you shoot at the VC?"

"No sir. When I rolled out of the truck, I stuck the barrel of my weapon in the mud and I was afraid to fire it."

"What time of day did this happen?"

After a long pause he mumbled, "A little after 1800."

"You stupid bastard! How often have you been warned about running the roads after 1700? Nothing is safe around here when the sun begins to go down. You keep forgetting that you're a REMF—not a grunt. Grunts know better than to do that kind of shit."

The young sergeant described how the VC had searched through the tall grass while he and his buddy pressed themselves into the mud. He didn't understand the language, but he could hear a female shouting orders to the searchers.

"Yeah, there's a local-force VC platoon with a woman leader, and she's supposed to be pretty tough. How'd you get away?"

The kid said that a nearby PF outpost had heard the shooting and had radioed province headquarters. Just when he thought that they were about to be captured, a reaction force showed up and the VC took off into the woods.

"I hope you learned a lesson."

"Yes sir. I ain't leaving this compound until I go home."

The next morning I got a frantic call to get to the TOC immediately. Wondering what was happening, I double-timed over to the new underground operations bunker they'd built after the rocket attack. Quickly, Major Clark briefed me; intelligence was reporting that the headquarters of the 9th VC Division was supposed to be south of Phu Cuong, near the river. Pointing to a spot on the map, Clark said that the intelligence people believed the division's command group and a signal detachment were there, protected by a VC reconnaissance company.

He didn't say where they had gotten the intelligence, but when he mentioned the signal detachment, I figured that it came from one of the army's secret "Radio Research" units. They had ground and airborne listening stations, which monitored radio frequencies and triangulated on suspicious transmitters. The details were hush-hush, but everyone knew what these stations did. It was pretty hard to conceal their unusual antennas.

After treating us to a hair-raising drive, our trucks pulled over to the edge of Highway 13, just south of An Son. The platoon leaders organized their people and Hien issued orders for the 2d Platoon to take the lead. The command group fell in behind the 2d Platoon, and the other two platoons brought up the rear. The platoon leaders put out point and flank security, and we began moving quickly down a narrow dirt road. None of us liked walking on a road, but we had several kilometers to cover before we could deploy and start our search for the 9th VC headquarters.

The 9th VC was one of the units we regularly fought. Officially it was a Viet Cong division, but it had a lot of NVA soldiers in it. It had three regiments, with three battalions each: the 95-C, the 271st, and the 272d VC Regiments. The 272d VC Regiment operated primarily in Binh Duong Province and was the one we ran into most often, but this was the first time we'd ever targeted the division headquarters itself.

We were making good time and we were about one kilometer from the river when Hien got a radio call and halted the company. G-2 was getting new intelligence: The 9th Division headquarters had moved a couple of kilometers south of where we were headed. We couldn't cross a large canal between us and the new area, so we were ordered to double-time it back up the road to meet the trucks.

We had moved quickly but cautiously, because we knew about recon troops and we weren't about to charge mindlessly into them. But since we were retracing the ground we had just covered, Hien told the platoon leaders to pull in their flank security teams so we could make better time.

We were walking as fast as we could and I was staring at the road in front of me when I not only heard the cracking of AK-47 bullets, but also felt the air splitting as they went by me. I made a swan dive and ended up in a deep ditch alongside the road. Immediately Glen and the others were in the ditch with me. We'd been ambushed in a spot we had walked over just minutes before!

Only the VC were firing, because everyone in the Recon Company was busy scrambling for cover. The bullets were snapping over the top of the ditch, and they appeared to be coming from a half-dozen AK-47s, which were opposite where our command group was huddled. I looked at Hien and saw that he had come to the same conclusion. If the VC had set up an ambush near a ditch, they probably had claymores positioned along the bank to blow us away. No words were spoken—we just jumped out of the ditch and charged the VC.

The army teaches something called "immediate-action drills." The immediate-action drill for an ambush is to charge it; firing everything you have. The rationale is that the enemy will have prepared the ambush site to catch you in the killing zone. The longer you stay in the killing zone, the more casualties you will take. The only thing to do is to attack the ambushers. The command group was caught in the middle of the killing zone, so we were the ones who had to attack to get out of it.

The platoons on either side of us began moving off the road to flank the VC, while six of us in the command group assaulted. As I moved along a barbed wire fence that ran directly into the VC's position, I had my M-16 on full automatic and was firing as fast as I could.

It seemed like longer, but it was over in about a minute. The VC were lying on the ground, and we charged into them before they could even change the magazines in their AKs. Crouching forward, I sprayed almost a full magazine into the three I could see. Hien and the others shot the ones that I couldn't see. None of the six VC survived, and amazingly, none of us were hit.

Apparently this was a "hasty" ambush, meaning that the VC hadn't had the time to properly set up and prepare the site. They hadn't put out claymores or set booby traps. The only thing they had were their AK-47s. They'd scraped out shallow prone positions and had covered themselves with leaves. We figured they hadn't triggered the ambush on our way down the road because our flankers had walked behind them. When we pulled in the flank security on our way back, the VC had their opportunity. They waited until they saw the command group's telltale radio antennas and then opened up. They weren't trying to wipe out the whole company, only the leaders. They figured on escaping through the trees and brush during the confusion. It hadn't worked quite the way they had planned.

After checking the bodies, I looked around for Glen and Collins —they were just then climbing out of the ditch. I walked back to the road and asked them why they hadn't assaulted with us. Holding up his 35mm camera, Glen grinned and said, "Shit, Bob; you were doing so good that I decided just to stay here and take pictures of the whole thing."

We were still talking off the adrenaline surge when I looked at Glen's breast pocket. He had a metal Vietnamese spoon sticking up, where one would normally put a pencil. I reached over, pulled it out, and held it in front of his face—it had an AK-47 hole in it. Stunned, he looked at it for a few seconds, and then we all began to check our uniforms. He had another hole in a floppy jacket pocket, and I had a graze mark along the front of my shirt. The VC couldn't have shot any closer and still missed. Not one of us had been wounded—but we all had come very close to buying it.

The trucks took us a couple of kilometers south to another dirt road. Again we unloaded and began moving toward the river. Division was pushing us to move quickly and didn't seem to care that we had just been ambushed doing exactly that. We cared, so we ignored Division and made certain that we wouldn't get surprised again.

This time the 1st Platoon was leading, and as its point team rounded a bend in the narrow dirt road, a VC machine gun opened up on it. The platoon deployed on either side of the road and the command group moved up to control the action. Peering through the bushes, I spotted where the fire was coming from—a small red brick house. I was trying to figure out where to move to get a shot at it with my LAW, when one of our M-60s began returning the VC's fire. The gunner had the target and was firing continuously into the hootch. I worried he'd burn up the barrel, but I was fascinated with the way the 7.62mm rounds were eating through the bricks. The bricks weren't the small solid type we used in the States; they were larger and hollow, like cinder blocks. Red dust and small chips were flying as the bullets punched into, and then through them. In a few seconds, I could see the interior of the hootch as the holes got bigger and bigger.

The machine gunner came to the end of his 100-round belt, and the firing stopped while he and his assistant loaded another. I motioned him to hold his fire, and we waited to see if we had gotten the VC. The hootch was silent, and Hien directed the 1st Platoon leader to send a squad to check it out. The soldiers cautiously moved in from the flank and tossed a grenade in the door. After a muffled blast, two of them rushed inside. A couple of seconds later they came out and waved that it was okay.

The rest of the platoon and the command group moved up to the house. Lying in the rubble were two dead NVA, an AK-47, and an RPD machine gun. First we had been hastily ambushed by six NVA, and then we'd been fired on by two more. It seemed likely that both groups had been part of the outer security ring of the 9th VC Division headquarters.

Hien reported what had happened to Division, while Glen and I chuckled about how the M-60 machine gun had just eaten big holes through the bricks. Two of the point men had been wounded, and Hien directed a couple of soldiers to help them back to Highway 13. We redistributed ammunition and resumed moving toward the river. We figured we were close and got the hell away from the road.

This area was lightly populated, with hootches every couple of hundred meters. It was mostly fruit trees and small garden plots bounded by narrow ditches filled with murky water. The bad guys were somewhere around, because there wasn't a civilian in sight. I got one of those eerie feelings—I could sense the VC's presence.

I hated this kind of area. The tree branches started about four feet above the ground, and the leaves were so thick that you couldn't see anything while you were on your feet. On the other hand, the VC could lie on the ground and see us coming a hundred meters away.

We had deployed from a column to a formation with two platoons forward and one to the rear. The lead platoons were spread out on a skirmish line so we could cover a larger area and not bunch up. Although it was harder to control, the formation allowed us the maximum amount of firepower forward, and it meant that the VC would need a large force to engage us all. If we did run into something, one or both of the wings could come around and flank the VC position. The trailing platoon could be put into the line wherever it was needed, or it could be used as a maneuver element.

The foliage overhead blocked the sunlight, and although it wasn't dark, it was gloomy as hell. Everyone sensed that we were about to make contact and were quietly picking their way ahead. It was one of those times when you really couldn't do anything, except pray that the VC missed you with their opening burst of fire.

It had to happen eventually—and it did. An RPG rocket exploded about 50 meters to my left, and the VC opened up with AKs on full automatic. I hit the ground behind a scrawny tree and began firing to the front. I couldn't see anything, but in a firefight you don't wait around for a target; you fire wherever the VC might be. The opening moments are critical, and you need to put out as many rounds as possible.

The trailing platoon crawled forward and moved into line with the rest of us. I knew that Hien was probably ordering the platoon on the right to maneuver around to hit Charlie on the flank. After a few seconds the firing tapered off, as people began to change magazines. I was on my third magazine when I heard the yell to cease fire. The VC had stopped shooting and our own people were sweeping across in front of us.

There was some shouting, and the troops around me began to get up from the ground. I got to my feet and joined Hien as he moved forward toward the maneuver element. We found one body and the RPG-7 he had been carrying. Apparently he had been too slow pulling back with his buddies. There were spent AK-47 shells lying all around, but nothing indicated that any of the other VC had been hit.

One of the platoon sergeants came up and said something to Hien, and he followed the man back to our original line. We found we had lost one man killed and another with his leg blown off just below the knee. The wounded man was one of my favorite soldiers. He always laughed and joked, and he made a point of waving whenever he saw me. It always upset me when we lost somebody, but there were some soldiers that I really hated to see hit. This guy was one of them.

I wanted to call in a Dust Off, but the damned trees were so thick that there was no way a chopper could land between where we were and Highway 13. Hien told the platoon leader to make a couple of litters and carry the dead and wounded men back to the highway. I had to stay with the company, so Division would have to send an ambulance from Lam Son to evacuate the man.

Once again we got ourselves organized and began moving toward the river. We had definitely been running into the 9th VC Recon Company. The three groups of VC had been too professional, too well armed, and too willing to take casualties to be local-force units. They had our number and were slowing us down.

About thirty minutes later, Division called and said that its intelligence was now reporting that the VC headquarters had crossed the Saigon River. (There was no question about it. They had to be intercepting the VC radio transmissions.) We pushed to the river bank but couldn't find any way to cross. When we told Division, it radioed for us to return to Highway 13 for pickup. There was nothing else we could do.

We covered the distance pretty quickly, since we were fairly certain that the 9th VC Recon had crossed the river with its division headquarters. When we got to the highway, we found our dead and wounded men still waiting for evacuation. It had been over an hour since we had called for an ambulance, and nothing had shown up. My wounded buddy had lost a lot of blood and had gone into shock. Instead of being light brown, his face was gray-yellow, and he was in bad shape. I checked the medics, but none had any blood expander, so there was little we could do except wait for transportation.

Hien was angry and chewed somebody's ass over the radio. I knelt down next to the man and tried to make him feel better. He reached out and took my hand in a weak grip and wouldn't let go. He tried to say something, but I put my finger to my lips so he wouldn't waste what little strength he had left.

Hien threw the handset on the ground and told a couple of soldiers to flag down the next vehicle headed north. Normally, the highway was heavily traveled, but this was the traditional siesta time and it was empty. A couple of motorcycles and scooters passed, but we couldn't use them. Finally, a three-wheel Lambretta came around the bend and the soldiers stopped it. They directed it to the side of the road, and we got ready to put the litter in the back.

Then I looked at my little friend and saw that his eyes were clouded and beginning to glaze over. I didn't have to check his pulse—he was dead. I pried his hand free of mine and walked over to some bushes where I could cry privately.

I was in a foul mood when we got back to Lam Son, and I marched straight to the TOC. I didn't say anything to the Americans, but I thoroughly reamed out the ARVN operations officers. Although it probably wasn't true, I blamed them for the death of my Recon buddy. (He probably would have died even if he had been evacuated in time.) I surprised both the ARVNs and the Americans with my extensive vocabulary of Vietnamese curse words. They didn't say anything, but I noticed that one of them kept glancing at the small skull and crossbones on my beret. I may have scared them, but I knew they were even more frightened of the Recon blaming them for the guy's death—fragging wasn't a solely American practice.

None of the Americans interfered, and when I was done, there was silence in the bunker. I started to walk out when I noticed Major Binh standing in the corner. I had only seen him a couple of times since I had filled in as his senior advisor almost a year before. He noticed the Vietnamese captain's insignia on the front of my camouflaged shirt and said, "I see you've been promoted to captain. I thought you said it would be several years before you would be promoted."

Giving me a sly look, he added, "Of course, your Uncle Westmoreland had nothing to do with it."

I just smiled.

I guess the army hadn't forgotten me, because the next day I received my orders. Since I'd gotten the note, I had been giving the assignment a lot of thought. I didn't really want to spend the time between my Vietnam tours wading around the Florida swamps, but there was prestige in being a Ranger instructor. It was the toughest training in the army, and I would get an extra $110 a month, hazardous duty pay for jumping out of airplanes.

I don't know why, but my orders seemed to catch everyone off-guard. I had been around so long that nobody figured I might actually pack up and go home. Major Clark seemed a little surprised when I gave him a copy of my orders and found I had less than three weeks left. With a blank look, he said that he guessed they had better find somebody to replace me.

Finding a replacement was a little difficult. The Americans could ram an advisor down ARVN's throat for most positions—but not in the case of the Recon and Recondos. General Thuan, Major Tu, and Lieutenant Hien had to agree with the selection. Some of our advisors were good combat troops, but they didn't get along very well with their Vietnamese counterparts. Others got along well, but they weren't experienced enough to keep up with the Recon. Finally, there were officers who were good on both counts, but they weren't about to volunteer for five or six months of our kind of combat.

I recommended one guy, and Clark and Sonstelie were reasonably satisfied with my selection, but the Vietnamese still hadn't given their approval. They were hoping that I'd extend my tour. (My wife probably would divorce me if I did.)

Finding a suitable NCO replacement was a lot easier. Chuck was due to leave about the same time, and we hadn't had any trouble finding someone for his position. The sergeant was a quiet, mild-mannered guy, but with good credentials. He was a little older than Glen and the others, but according to what we had found out, he could hump a rucksack as well as anyone.

The day after I gave Major Clark my orders, he informed me that I would be grounded the following Friday. I argued that that was more time than I needed, so we finally agreed on the fourteenth, ten days before I had to report to Saigon.

During the next week or so, the Recon didn't get into any heavy action. The Cav Squadron did corner some local-force VC in a laterite quarry, but when we got there the situation was under control. It was the only time I ever saw a flamethrower used in Vietnam, and except for a few minor glitches, it worked well.

For some reason the VC decided to hide in a cave rather than to try to slip away. Caves and tunnels weren't unusual, because the VC had them everywhere, but this one hadn't been dug by the Viet Cong. It had been made by stonecutters taking blocks of laterite out of the side of the hill for tombstones—a booming business in Vietnam. The VC squad had ducked into the cave, and then it found that there was no way out. The armored personnel carriers just surrounded the entrance and ordered the VC to surrender.

The Recon sat back and relaxed while the Cav commander yelled to the VC to come out. If the trapped Viet Cong were answering the ARVN captain, I couldn't hear them. The Cav troops screwed around for a few minutes, and then a soldier went to the back of one of the tracks and hauled out a flamethrower. After some more calls to surrender and some tinkering with the flamethrower, the soldier poured a stream of burning gas around the entrance to the cave. That got the VC's attention, because five of them came charging out, firing their weapons. That was a dumb move, because the .50-calibers opened up and blasted them. They went down in a heap, and a squad of ARVN soldiers moved up to get their weapons and to see if there were any survivors—there weren't.

The man with the flamethrower moved closer and shot another stream of fire at the entrance. In an instant two more VC broke out of the cave and were brought down by the machine guns. It reminded me of newsreels showing the Marines rooting out Japanese soldiers in the Pacific.

After filling the cave with flame, we walked over to the VC bodies to see what the Cav had gotten. Everyone was dead, except a woman who had taken a half-inch round through the pelvis. There was a big hole in her side, and her intestines were sliding out on the ground. Her hair was singed and still smoking from the fire. She had a hard time breathing, but she was snarling more dirty words than I had learned in the year I'd spent in Vietnam. She couldn't move her legs, but she kept striking out at us with her arms. She was one tough lady.

There wasn't much anyone could do and she died a few minutes later. The soldiers pulled off her ammunition pouch and searched her clothing. In one of her pockets they found a bundle of papers wrapped in plastic. Included was a certificate for killing more than ten Americans. Hien recognized the name on it and said that she was the platoon leader of a particularly nasty group of local-force Viet Cong—the same group that had ambushed the young sergeant on Highway 13.

The next night I was racked out in my hootch when one of the perimeter guards came in and said that his people had somebody in the wire. I grabbed my web gear and rifle and followed him to the berm. It was pitch black, but one of the guards handed me a Starlight scope, which I used to scan the minefield. I could just make out a shape that looked like a man hiding in the tall grass. I whispered to the guards to pass the word for everyone to open up when I fired a hand-held illumination flare.

I was just getting ready to launch the flare when Colonel Sakas appeared at my elbow. I stopped and told him that there was somebody, probably a VC sapper, in the wire. There were likely to be others, but I had only spotted one. I started to fire the illumination, but Sakas stopped me. "Are you certain it's not a lost ARVN patrol?"

"You must be kidding. With all the lights in the compound, how could an ARVN patrol get lost and wander into our wire?"

"You're probably right, but I want to check with the TOC before we do anything."

It's hard to scream when you have to whisper, but I did my best. "Colonel, you must be out of your fucking mind! Friendly patrols don't crawl into other people's minefields!"

"Watch your mouth, Captain! I don't want you to do anything until I have checked with the TOC."

"Okay, Colonel—but it's all on your head now."

Apparently, with all our screwing around, the VC realized that they'd been spotted. One of the guards yelled, and I immediately slapped the detonator cap, firing the flare. In the glow we saw a half-dozen figures running for the far tree line, and the guards opened up. The mortars started firing illumination rounds, but the grass was so thick that nobody could see where the VC had gone. One of the flares drifted down into the minefield, setting the grass on fire. (I had repeatedly asked for permission to burn this very same grass, but had gotten nowhere.) The fire spread quickly and began setting off claymores. The explosions threw sparks and burning embers, starting more fires. Undoubtedly the VC had gotten away, but if any had been foolish enough to stay behind, the fire and the exploding claymores would have taken care of them.

I watched the whole thing for about fifteen minutes and then walked back to my hootch. With all the excitement, I was the only man in the compound who wasn't along the berm or in a bunker, but I didn't give a damn. I only hoped that the men wouldn't wake me up when they finally decided to go back to bed. I was pissed at Sakas, but I was also beat, so I dropped off to sleep pretty quickly.

About 0400 in the morning, I heard an explosion and then the sound of the alert siren. I grabbed my gear and ran to the berm, expecting incoming mortars and a VC attack. As I got closer, I saw the perimeter guards standing outside their bunkers, peering over the top of the berm. I grabbed one of them and asked him what was happening. Earlier, the fire had ignited the foo-gas barrels, and they had been burning for several hours. When the flames got to the bottom of the one of the gas drums, the heat set off the TNT. There had been an explosion, and in a cloud of smoke and white phosphorus, the oil drum had been blown 50 feet into the air. He pointed at a couple more foo-gas barrels, saying that they should be going up soon.

I watched for a few minutes, and then the next one exploded. It wasn't as spectacular as a full barrel of jellied gasoline would have been, but it was still kind of neat. After the third one blew, I walked back to my hootch, swearing that I would tell Colonel Sonstelie about Sakas screwing around with the sappers, and about how I had tried to get him to let me burn off the grass. I also swore that I wouldn't replace the foo-gas. They could get some other sucker to do that.

On the fourteenth, the Recon choppered into the Hook without me. I had asked Clark for one more day in the bush, but neither he nor Sonstelie would agree. I was grounded, and that was the end of my fighting for this tour. I sat in the TOC and monitored the insertion on the radio. The LZ was cold, and Tony reported that they were moving out to search for the VC. I hung around for an hour or so and then wandered back to the G-2 office.

I was lounging in a chair, trying to kill some time, when a radio operator stuck his head in the door and said the Recon had gotten into contact. I raced to the TOC and crowded up to the radio to hear what was happening. The Recon was engaging an estimated platoon of VC, and our new sergeant was down with a bullet in the leg. Although they had other wounded besides him, they weren't having as much trouble with the VC as they were with the RAG boats on the Saigon River. The boats had received some Viet Cong fire and were returning it with 20mm guns. The Recon was pinned down, with the VC shooting from one side and the RAG boats firing from the other.

I'd had my share of frustrations in Vietnam, but none had been worse than this. My people were pinned down and taking casualties, while I was sitting safely in the TOC. I raced back to the G-2 office and told Major Clark that I had to get in to help the company. I was going to get the admin chopper to land me south of the Recon, and then I'd work my way up to them. I ran to the map to show him where I was going and what route I would take.

It was like talking to a stone wall. He just shook his head and said that my war was over. He'd just lost one advisor and wasn't going to lose another, particularly one who was packing to go home. The more I raged, the more adamant he became—I was done. He picked up the telephone and called the chopper pad. "Under no circumstances are you to allow Captain Parrish on a helicopter for anything."

I slumped into a chair and fumed. Every time I thought up a new argument, Clark told me no. "Now you can understand how we feel when you are out there in a firefight. We crowd around the radio, waiting to see if you guys are going to be all right."

Tony and Glen finally got the RAGs to cease fire, and they were able to medevac the wounded sergeant. They'd lost four men and killed eight of the VC before Charlie melted away. I met them at the airstrip when they returned and felt like a damned REMF.

I spent the next few days getting my stuff together and registering the war trophies I was going to take back to the States. Then on the last Sunday before I was to leave, Hooker asked me if I'd like to go with him on an aerial recon. I was so tired of sitting around the compound that I jumped at the chance. I knew I'd get my ass chewed if anyone found out, but the last week had dragged by so slowly that I had to do something to maintain my sanity.

It was a beautiful day, and the ground below was surprisingly peaceful. Normally you could see smoke from airstrikes or artillery somewhere on the horizon, but today there didn't seem to be much happening. I asked Hooker to fly over Phu Hoa Dong so I could take one last look, and then we crossed the river into the Iron Triangle. As we always did, every time we spotted something suspicious, Hooker took the plane in low and circled the spot.

I thought I saw something in the bushes, so Hooker came around and we flew back over it. Sure enough, there were a couple of camouflaged hootches and a small campfire. I stuck my M-79 grenade launcher out the window and Hooker kicked the Bird Dog into a tight turn.

When I had first tried firing a 40mm grenade from an airplane, I had problems. The motion of the plane caused the round to tumble, and if it hit on its side or base, it wouldn't detonate. After a little experimenting, we found that if Hooker held a tight turn and I aimed at the center of the whirling ground, the round usually exploded. A helicopter gunship didn't have that problem, because it fired its grenades in the same direction it flew. Shooting out the side of a fast-moving plane caused the tumbling. The grenade wouldn't tumble if I shot to the rear, but Hooker vetoed that idea because he was afraid I'd shoot a hole in the plane's tail. I blasted the hootches with a half-dozen rounds, but we couldn't tell if anyone was down there. I sprayed a full magazine from my CAR-15, and then we headed off to see what else we could find.

The air rushing in the window was almost like sitting in front of an air conditioner, and I felt good. I was becoming a little sleepy when Hooker told me to switch my radio over to the emergency frequency. When I did, I heard another Bird Dog broadcasting a Mayday. His plane had been hit by groundfire, and he was taking evasive action. In a panic-stricken voice, he said that the guy in the back seat had been badly wounded.

For about thirty seconds, the pilot kept repeating "Mayday! Mayday!"—and then there was silence. I had just about concluded that he had gone down when he came back on the air. He'd gotten away, but his observer was in bad shape. The man had taken a round or a chunk of metal in the groin and was losing blood fast. I could hear the pilot choking as he described how the blood was flooding the floor of the airplane. It was tough to sit there and listen to him, knowing that there was nothing anyone could do to help. After a few more minutes, in a more collected but resigned voice, he broadcast that his passenger was dead.

As I sat there imagining what was going on in the other plane, I suddenly realized how stupid I was being. I had less than five days to go in country, and I was screwing around over the Iron Triangle in a fragile little airplane. I punched the intercom button and told Hooker to get me the hell back to Lam Son.

During the next couple of days, I had to attend my farewell parties—the Recon and Recondo companies, the division advisory team, and my own team. I was a very unhappy guest of honor, and the booze didn't prevent a lot of long faces. I really hated to leave.

I thought that the Recon party was the last hard time that I would have in Vietnam, but I was very wrong. Two days before I was to leave, I was walking through the compound parking lot on my way to grab a cup of coffee. I was just passing a truck when I heard and felt a massive explosion. In an instant I was on the ground under the truck, while dirt and pieces of steel fell from the sky.

I had no idea what had happened, but it was the biggest explosion that I had ever been close to. At first I thought an aircraft must have dropped a thousand-pound bomb on the compound, but I hadn't seen or heard a plane. When the debris stopped falling, I crawled out from under the truck and saw a giant cloud of smoke and dirt rising from the area where my Recondos lived. *The damned captured VC ammunition must have blown up!*

Tony ran up and we jumped into the nearest jeep. I almost rolled the jeep several times, racing through the buildings to get to the Recondo compound. We roared around a hootch and were met with a scene of absolute destruction. Directly to our front was what was left of a large truck and shattered trees. The side of the Recondo barracks was blown away, and the inside was in shambles. Everywhere we looked, the ground was covered with dirt clods, unexploded ammunition, and blackened chunks of something. It was clear what had happened. An ARVN work detail had been unloading a truckload of captured ammunition, rockets, and mortar rounds. Somebody had dropped something, and the whole truck had blown up.

Stunned, I climbed out of the jeep and began to walk toward the Recondo barracks, dreading what I knew I would find. Something rubbery caught in the cleats of my jungle boots, and I reached down to pull it off—it was a piece of human flesh. As I threw it to the ground, I saw a human hand lying in the dirt. I looked around and there were body parts everywhere, some identifiable, others just blackened chunks of meat. Hanging on the barbed wire fence next to me was part of a human torso, without arms, legs, or a head. Nearby was a boot with a foot still in it.

An ambulance and some other vehicles screeched to a halt behind my jeep, and everyone climbed out in a daze. Tony and I picked our way through the carnage, searching the wreckage. The others regained their senses and began following us. Somebody yelled in English to watch out for unexploded ordnance, but Tony and I just ignored him. These were our troops, and we were going to see if we could save any of them. We needn't have bothered. In the barracks wreckage we found only broken and shredded bodies—everyone was dead.

I lost track of the time, but finally we walked back to the jeep. The only thing anyone could do now was literally to pick up the pieces of the Recondos who had been in the blast area.

I don't remember much of what happened during the next two days. If I had been a drinking man, I'd have stayed drunk most of the time. However, I wasn't, so I just tried to shut everything out. After the death of my Recondos, I only wanted to get the hell out of Vietnam.

Major Clark asked me if there was anything I'd like him to put in my efficiency report. I told him to write whatever he wanted, but to please add a line on what we had been doing. I didn't want anyone to think that I had been sitting on my ass for the past year. When he finished the form, he pointed to the statement I'd asked for:

During the past four months Captain Parrish has participated in 69 assorted combat operations, of which 35 made contact with a squad or larger size enemy force . . . his unit enjoyed a 15 to 1 kill ratio.

The team took me to Saigon, but even that was typical of my tour. We had to take cover in a ditch because of a firefight between an ARVN battalion and some VC. They were just off the edge of the road, and bullets were cracking across Highway 13. When there was a pause in the shooting, we jumped back in the jeep and Glen raced through the battle.

Processing in was comparatively quick and painless, because the Saigon warriors knew better than to mess with grunts just in from the bush. On the morning of 25 April 1968, a bus took us to Tan Son Nhut, and after a couple of hours of sitting around, we heard the loudspeaker announce the boarding of the chartered Continental Airlines flight to Travis Air Force Base. We filed on the airplane and almost everyone cheered as the plane took off. For some reason I didn't feel like cheering, so I just sat in my seat and watched the Vietnamese countryside grow smaller and smaller—until there was nothing left but the South China Sea.

I lost track of the Americans with whom I served. Glen Septer was assigned to the army's Mountain Ranger Camp, and then he went back to Vietnam for another tour. I hope he and the others survived and are well.

George was last seen being chased around by a male monkey after I gave her to the 7th Regiment's advisor team.

In 1973 I ran into Captain Hung at Fort Benning, Georgia, where he was attending the U.S. Army's Infantry Officer Advanced Course. I brought him home to meet my family, and we spent the evening going over old times. My wife listened and smiled, but she didn't understand any of it because the entire conversation was in Vietnamese. He told me that after the explosion at the barracks, the G-2 Recondos received forty or fifty replacements, and the company was reconstituted. The 5th Recon and the G-2 Recondos continued to fight and lose people, both Vietnamese and Americans. Finally in April 1972, the companies and all the rest of the men I had known were wiped out while fighting NVA tanks in An Loc, the capital of

Binh Long Province. Hung went back to Vietnam after the training course and disappeared in the war.

The 3d Battalion was destroyed in the final attack on Saigon in 1975, not long after the U.S. Congress refused to approve Vietnam's request for emergency ammunition.

In 1970, after serving as an instructor in the Ranger Department and attending the Infantry Officer Advanced Course, I returned to the III Corps area. I spent eight months in the bush, commanding a rifle company in the 1st Cavalry Division, until I was medically evacuated back to the States.

According to my notes and after-action reports, I estimate the 3d Battalion, the 5th Recon Company, and the G-2 Recondos lost about 300 men killed and wounded in combat during the year I was with them. That was a very high casualty rate for *a bunch of ARVNs who wouldn't fight.*

MEN AT WAR
The battles. The blood.
The way it was.

KNIGHTS OF THE BLACK CROSS
Hitler's Panzerwaffe and its leaders
by Bryan Perrett
_____ 91130-0 $4.50 U.S.

SHINANO!
The Sinking of Japan's Secret Supership
by Capt. Joseph F. Enright, USN, with James W. Ryan
_____ 90967-5 $3.95 U.S.   _____ 90968-3 $4.95 Can.

KOMMANDO
German Special Forces of World War II
by James Lucas
_____ 90497-5 $4.95 U.S.

BUSHMASTERS
America's Jungle Warriors of World War II
by Anthony Arthur
_____ 91358-3 $4.95 U.S.   _____ 91359-1 $5.95 Can.

FORTRESS WITHOUT A ROOF
The Allied Bombing of The Third Reich
by Wilber H. Morrison
_____ 90179-8 $4.95 U.S.   _____ 90180-1 $5.95 Can.

GREAT BATTLES OF HISTORY
FROM ST. MARTIN'S PAPERBACKS

OPERATION TORCH
The Allied Gamble to Invade North Africa
by William B. Breuer
———— 90125-9 $3.95 U.S.   ———— 90126-7 $4.95 Can.

STORMING HITLER'S RHINE
The Allied Assault: February–March 1945
by William B. Breuer
———— 90335-9 $4.95 U.S.   ———— 90336-7 $5.95 Can.

THE LAST BATTLE STATION
The Saga of the *U.S.S. Houston*
by Duane Shultz
———— 90222-0 $4.95 U.S.   ———— 90223-9 $6.25 Can.

DELIVERANCE AT LOS BAÑOS
by Anthony Arthur
———— 90346-4 $3.95 U.S.   ———— 90347-2 $4.95 Can.

THE LAST ENEMY
by Richard Hillary
———— 90215-8 $3.95 U.S.